Praise for

· THE NATION OF PLANTS ·

"A renowned scientist delivers a simple yet urgent call to action on behalf of Earth's multitude of plants . . . [a] powerful book . . . Mancuso concludes his elegant and cogent argument with straightforward advice accessible to anyone . . . Insightful and arresting, this book offers an achievable road map to a more 'radiant future.'"

—*Kirkus Reviews*

"In this brief book, Stefano Mancuso offers what may be the most original solution to the troubling age of humans. What if it were plants, rather than humans, who wrote a constitution for Earthly survival? Mancuso's innovative manifesto is a set of principles for living according to the botanical world. He imagines a new political order based not on the survival of the fittest, but rather on life in community, mutual aid, freedom from borders, and sovereignty for all living beings. In this engaging read, the plant philosopher pushes readers to see how much our survival depends on the well-being of the Nation of Plants—and gives us a radical guide to living according to the rules of life's unsung heroes."

—Elizabeth Hennessy, author of *On the Backs of Tortoises: Darwin, the Galapagos, and the Fate of an Evolutionary Eden*

"*The Nation of Plants* unveils the long-term relationship between plants and people and explores the rights of all living things. It is a call for cooperation in a world facing persistent environmental degradation. It is a call for our mutual survival."

—Lauren E. Oakes, author of *In Search of the Canary Tree*

"A fantastic and necessary read for any plant enthusiast or environmental activist, *The Nation of Plants* is not merely a missive on the perils of climate change. Rather, the book begins from the whimsical perspective of plants, then weaves scientific fact with historical examples in a moving and inspiring call to action. Apart from the initial address, Mancuso's concrete approach is far from fantastical. *The Nation of Plants* is moving and informative, balancing a love for all things botanical with a passion for listening to and considering the needs of our plant brethren."

—Jessica Roux, author of *Floriography: An Illustrated Guide to the Victorian Language of Flowers*

"In his new book, *The Nation of Plants,* Stefano Mancuso expresses his awe for plants by asking a unique question: What if our constitution were rewritten by plants? What would be the fundamental laws if the Earth were governed by plants rather than people? Mancuso answers this question by masterfully and thoughtfully linking the stories of people, plants, and plant science. A must-read for anyone who is interested in the historical interactions between people and plants."

—Valerie Trouet, author of *Tree Story: The History of the World Written in Rings*

· THE NATION OF PLANTS ·

ALSO BY
STEFANO MANCUSO

The Incredible Journey of Plants

The Revolutionary Genius of Plants: A New Understanding of Plant Intelligence and Behavior

Brilliant Green: The Surprising History and Science of Plant Intelligence

THE NATION OF PLANTS

Stefano Mancuso

TRANSLATED FROM THE ITALIAN
BY GREGORY CONTI

OTHER PRESS
New York

Originally published in Italian as *La nazione delle piante*
in 2019 by Editori Laterza, Rome.

This book was translated thanks to a grant awarded by the
Italian Ministry of Foreign Affairs and International Cooperation.

Production editor: Yvonne E. Cárdenas
Text designer: Jennifer Daddio / Bookmark Design & Media Inc.
This book was set in Horley Old Style by
Alpha Design & Composition of Pittsfield, NH.

1 3 5 7 9 10 8 6 4 2

 Printed in the United States of America on acid-free paper.
For information write to Other Press LLC,
267 Fifth Avenue, 6th Floor, New York, NY 10016.
Or visit our Web site: www.otherpress.com

Library of Congress Cataloging-in-Publication Data
Names: Mancuso, Stefano, author. | Conti, Gregory, 1952- translator.
Title: The nation of plants / Stefano Mancuso, translated from
the Italian by Gregory Conti.
Other titles: Nazione delle piante. English
Description: New York : Other Press, 2021. | Includes
bibliographical references.
Identifiers: LCCN 2020030307 (print) | LCCN 2020030308 (ebook) |
ISBN 9781635420999 (hardcover) | ISBN 9781635421002 (ebook)
Subjects: LCSH: Plants--Law and legislation. | Human-plant relationships.
Classification: LCC QK50 .M3613 2021 (print) | LCC QK50 (ebook) |
DDC 580—dc23
LC record available at https://lccn.loc.gov/2020030307
LC ebook record available at https://lccn.loc.gov/2020030308

• CONTENTS •

PROLOGUE

More than fifty years ago, on Christmas Eve 1968, the Apollo 8 mission became the first to carry a human crew in orbit around the moon. William Anders, Frank Borman, and James Lovell were the first lucky mortals to observe the dark side of our satellite and experience the enchanting spectacle of the rising Earth. In the course of that mission, during one of their ten moon orbits, William Anders took a photo that would become famous, rightfully earning its place among the icons of recent human history: the dawn of the Earth seen from the moon. Every one of us, at one time or another, has seen it reproduced. It shows the terrestrial globe, partially dark in its lower part, with the South Pole on the top left and South America in the center of the picture, rising above the lunar horizon. A blue and green world, with white clouds woven delicately across its entire surface. That picture, dubbed by its author *Earthrise* and catalogued by NASA with the less poetic entry number AS8-14-2383HR, changed forever our idea of Earth, revealing for us a planet of majestic beauty, but also fragile and delicate. A colorful island of life in an otherwise empty and dark universe.

A planet green with vegetation, white with clouds, and blue with water. These three colors are the signature of our planet but, for one reason or another, they would not exist without plants. Plants are what make Earth the planet we know. Without them, our planet would very much resemble the images we have of Mars or Venus: a sterile ball of rock.

Yet of these beings that make up nearly the totality of living things, which have literally formed our planet, and on which all animals depend—humans, obviously, included—we know extremely little, almost nothing. This is an enormous problem, one that impedes us from understanding how important plants are for life on Earth and for our personal, immediate, survival. By perceiving plants as being much closer to the inorganic world than to the fullness of life, we commit a fundamental error of perspective, which could cost us dearly. In an effort to make up for the scarce awareness and esteem that we have for the vegetable kingdom, given that we humans comprehend only human categories, this book treats plants as though they were part of a nation, or a community, of individuals, who have common origins, customs, histories, organizations, and goals: the Nation of Plants.

Looking at plants in the same way we look at a human nation leads to some surprising results. The Nation of Plants, with its green, white, and blue tricolor flag (they are the colors of our planet and they depend on the presence of plants), is the most populous, important, and extensive nation on Earth (trees alone number more than three trillion).[1] Comprising every single vegetable being on the planet, the Nation of Plants is the nation on which every other living organism depends. So you thought that the superpowers were the true masters of the Earth, or you believed that you depended on the markets of the United States, China, and the European Union? Well, you were wrong. The Nation of Plants is the only true and eternal planetary power. Without plants, animals would not exist; life itself, perhaps, would not exist on our planet, and if it did, it would be something terribly different. Thanks to photosynthesis, plants produce all the free oxygen present on the planet and all the chemical energy consumed by other living beings. We exist thanks to plants, and we will continue to be able to exist only in their company. It behooves us to keep this idea clear at all times.

Even if they behave as though they were, humans are not the masters of the Earth, but only one of the most unpleasant and irksome residents in the condominium. From the moment of their arrival, about 300,000 years ago—nothing compared to the history of life on our planet, which goes back to 3.8 billion years ago—humans have succeeded in the challenging enterprise of changing the conditions of the planet so drastically as to make it a dangerous place for their own survival. The causes of this reckless behavior lie partly in humans' inherently predatory nature and partly, I believe, in our total incomprehension of the rules that govern the existence of a community of living beings. The last to arrive on the planet, we behave like children who wreak havoc, unaware of the value and significance of the things they are playing with.

I have imagined that plants, like attentive parents, have come to our aid once again, after making it possible for us to live and realizing our incapacity to develop autonomously, by giving us rules—in reality, their very own constitution—to use as a guidebook for the survival of our species.

This is just what the book you now have in hand is about: the eight fundamental pillars on which the life of plants rests. One more than the seven pillars of wisdom of T. E. Lawrence (the famous Lawrence of Arabia), but with no pretense to wisdom, just to plain convenience.

Imagining a constitution written by plants, for which I serve as a go-between with our world, is the playful exercise that has given birth to this book. A constitution written by plants, and in the place of plants, by someone who knows nothing about legal matters. My brother, who on the contrary is an erudite super-magistrate, warned me immediately about the risks I was taking by playing with sacred texts and advised me to forget about it. Like all good brothers, I did not listen to him, so now all

I can do is hope in the clemency of the court for the inevitable inaccuracies that I have managed to stick into the few articles of the Constitution of the Nation of Plants.

It is a short constitution; based on the general principles that regulate the common life of plants, it establishes norms applicable to all living beings. Humans, in fact, are not the center of the universe, but just one of the many million species that by populating the planet form the community of the living. This community is the subject of the vegetable constitution; not a single species or a few groups of species, but all of life taken together. Compared to our constitutions, which place humans at the center of the entire juridical reality, in conformity with an anthropocentricism that reduces to *things* all that is not human, plants offer us a revolution. As in one of those sentences where it is enough to change the tone or the accent on a single word to change the overall meaning into its diametrical opposite, so the constitution of plants, by shifting the accent from a single species to the whole community, helps us to understand the rules that govern life.

In the pages that follow, you'll find the articles of the Constitution of the Nation of Plants, just as they were suggested to me by plants themselves in my by-now-multidecade familiarity with these dear fellow travelers. Each article is accompanied by a brief explanation that should help to clarify its significance. Enjoy.

Address to the United Nations General Assembly by the Representative of the Nation of Plants

Mr. President, Mr. Secretary General, honored guests, and distinguished delegates:

I am here today representing the Nation of Plants to direct to this noble assembly an appeal that can no longer be postponed. The indiscriminate use of the resources of our planet, the increasing pollution of its atmosphere, and the resulting change in our climate are the most serious threat that humanity has ever faced in the course of its very brief existence. I make this appeal, therefore, to each and every one of you and to the nations you represent, so that you will begin to modify your behavior, before the consequences of your conduct become fatal. If you do not change right away, the damage for people and for all the natural systems that sustain you will be irreparable.

Today, for the first time, our nation, the oldest and most populous on Earth, asks for the floor and it speaks to you, beseeching you to listen to us and to consider our words with attention and sagacity. We have sustained animal life, including yours, ever since the beginning. The planet that we inhabit is alive because we are here. Water, oxygen, the climate depend on us. We are the engine of life. Be conscious of that.

Over time, you have learned to use us. We are the basis of your food chain: everything you eat comes from us. Your most important sources of energy come from us. Your medical care depends on us. We supply you with building materials, fabrics, colors, beverages, beauty, health, and endless other benefits. You have learned very well how to use what we produce. But now the time has come for you to begin also to use what we can teach you.

In China, in the area of Beijing, you are constructing a single urban agglomerate that will soon host 130 million inhabitants. Within thirty years from today, more than 70 percent of the human population will be concentrated in urban areas. Seventy percent of carbon dioxide, which is the main cause of global warming, is produced in cities, from where it is dispersed into the atmosphere. Unfortunately, our great forests are only able to absorb 40 percent of it. Even though the arboreal population of our nation numbers three billion individuals, we are still too few and too distant from the places where you produce carbon dioxide for us to be able to help you effectively.

Use us better immediately! Cover your cities with plants, not just in parks, flower beds, roadsides, and on gardens and terraces, but by wrapping every possible surface in plants! We adapt much better and much faster than you do. In a very short time, we learned to absorb carbon dioxide in those environments where it is present in the highest concentrations, like your cities. If we are closer to the places of production, we absorb much, much more . . . and we use it to grow. Transform your cities into urban jungles, and the benefits you will receive will be incalculable.

In 2017, during a seminar at the Pontifical Academy of Sciences, one of your leaders whom we most admire, Pope Francis I, said: "I wonder if this piecemeal third world war that we are living through is not actually becoming a world war for water." He was right: 90 percent of the wars that afflict your bellicose species today are happening along the so-called drought line, in areas where there's less than a hundred millimeters (four inches) of rain per year—the limit under which you are not able to cultivate us. A billion people do not have access to secure sources of drinking water, and another four billion suffer from a scarcity of water for at least one month per year. Drought and soil aridity are causing you problems that you are unable to manage. Because of the drying up of water resources, desertification, reduced access to fertile lands, and famines, hundreds of millions of members of your species do not survive.

How can you imagine a future in these conditions? Yet you keep on consuming 70 percent of the planet's available drinking water for agriculture. Look around, two-thirds of the globe are submerged in salt water. Let me say it again: learn from us. We know how to live on salt water alone. Our halophyte sisters evolved for this reason. Use our knowledge, and you can transform the oceans into an immense reservoir that you can draw on to feed the planet, without consuming the drinking water that is so necessary for you.

And then energy. Some 80 percent of your supply derives from fossil resources: petroleum, coal, natural gas. Fossil fuels created by our nation over millions of years, which you are exhausting and whose consumption is the principal cause of the climate change now in progress. You have to find more sustainable models of development for the future of your species. Current renewable energy sources are a possible solution, but they are costly and probably not as clean as needed. We can assure you that obtaining clean and potentially infinite energy is possible. Imitate our photosynthesis: a process that allows us to transform sunlight, water, and carbon dioxide into sugars and oxygen. Artificial photosynthesis would resolve all of your energy problems. Its possible applications would be endless, and they would invert the normal process of producing waste. You can replace polluting systems that produce energy with a system that, by producing energy, cleans the air.

You are animals, and as such preying on and killing other living beings is your nature. We understand that. But there is a difference between preying on and destroying, and your actions are rapidly bringing on a mass extinction. Hundreds of thousands of species are threatened by your behavior; almost two-thousand five hundred square miles of forest are disappearing every month,[2] and with them, the ecosystem that ensures the survival of countless organisms. Do not be surprised if your species too, humanity, ends up suffering the consequences. Some of you have already been affected. The ever-higher numbers of viruses and microorganisms that are passing from other animals to people are one of the most direct consequences of the alteration of natural ecosystems. Sixty percent of the new infectious diseases discovered on a global level are zoonotic, and of the thirty new human pathogenic agents discovered in the last three decades, 75 percent have had animal origins. Ebola, SARS CoV-2 (aka the novel coronavirus), and HIV are just a few of them. We know and you know that your behaviors are the causes of your problems.

Life is a complex and unpredictable network of relationships, in which each living species is but a simple node. Every species depends on the network staying as intact as possible. The systematic elimination of organisms and environments is destroying the network that allows you to survive. The coronavirus pandemic that has struck you is like a small tremor of this network. Be wise and take remedial actions, or the next shock could have more dramatic consequences. In these months of confinement and quarantine, you have lived for a brief period of your lives like plants. Unable to move from the places where you live, remaining stationary inside your houses, you have had a chance to better appreciate the environment that surrounds you. You have learned not to waste resources that your life depends on. You have communicated more often and better with your neighbors. You have come to understand the importance of community. In sum, for a brief period you have come closer to the life of plants. Do not forget that.

If you would dedicate to plant research just one-tenth of what you spend on technology research, you would be saved. Up to now, you have preferred to do otherwise. Change! Now! We can teach you how to obtain energy from the sun; how to depollute our only planet, returning it to its ancient wonder; how to create organizations that are more democratic, decentralized, strong, with which to face a future you are now afraid of. In 2060, there will be ten billion of you, and you are asking yourselves how our planet will be able to maintain you all. Stop worrying! That is not the problem. Stop considering the other members of your species as rivals, as consumers of resources on the way to extinction. Take inspiration from us, reason and organize yourselves as we do, and three billion more human beings will become an enormous resource. Because every individual is a resource, an opportunity, a potential contributor to the solution of our collective problems. Open up your communities. Be cooperative. Closed borders do not protect your wealth; they only make you poorer.

Honorable Secretary General, dear delegates: we have asked for the floor because we believe it is the duty of our ancient nation to help you today, as always in the past. But we are not so used to talking. We are quiet by nature and you animals are so restless . . . always ready to run away from problems rather than resolving them. You are wearying, impulsive, proud. You prefer speed over contemplation, ephemeral power over the glory of life. The number of your shortcomings would not allow for excuses, but you are still a very young and inexpert species that knows how to learn fast. Do not be hardheaded. Do not persevere in error.

Learn from those who have more experience than you, and you will have a radiant future. We leave you as a gift our constitution, with the wish that it may help you to find the road to a long and happy cohabitation with us and our marvelous planet. Take care of yourselves.

I now return to the joys of my community.

The Constitution of the Nation of Plants

ART. 1 The Earth shall be the common home of life. Sovereignty shall pertain to every living being.

ART. 2 The Nation of Plants shall recognize and protect the inviolable rights of natural communities as societies based on the relationships among the organisms that compose them.

ART. 3 The Nation of Plants shall not recognize animal hierarchies, which are founded on command centers and centralized functions, and shall foster diffuse and decentralized vegetable democracies.

ART. 4 The Nation of Plants shall universally respect the rights of the currently living and those of future generations.

ART. 5 The Nation of Plants shall guarantee the right to clean water, soil, and atmosphere.

ART. 6 The consumption of any resource that cannot be reconstituted for future generations of living beings shall be prohibited.

ART. 7 The Nation of Plants shall not have borders. Every living being shall be free to travel, move, and live there without limitation.

ART. 8 The Nation of Plants shall recognize and foster mutual aid among natural communities of living beings as an instrument of coexistence and progress.

• ARTICLE 1 •

THE EARTH SHALL BE THE
COMMON HOME
OF LIFE. SOVEREIGNTY
SHALL PERTAIN
TO EVERY
LIVING BEING.

A surface area of 197 million square miles; almost 2,640 billion cubic miles of volume; a mass of 1.317×10^{25} pounds. These are the dimensions of our common home. At first glance, it might seem enormous. But it's not. When we compare its dimensions with those of other celestial bodies near to us, the sun, for example, whose volume is more than 1,300,000 times greater than that of the Earth, our home is shown to be what it really is, a small planet. But with some special qualities. It is, in fact, the only known place in the universe that has developed life. More than that, it is the only one on which life seems to *prosper.* What makes our planet special is not its size but its life.

The uniqueness of the Earth, the lack of credible alternatives capable of hosting life—

regardless of all that is commonly said about the possibilities of "terraforming" Mars or other improbable celestial bodies—makes it so that the entire planet should be considered a common good, inviolable, cared for and looked after as befits the only possible home for life. A home that also happens to be very fragile: limited to a thin surface layer that goes, more or less, from 33,000 feet below sea level to 33,000 feet above it; a total width of twelve miles that encloses the only place in the universe—as far as we know—where life exists.

A lot of people are convinced that the universe is full of life. Methodical calculations tell us of a universe more crowded than the Tokyo subway at rush hour. Perhaps. I wouldn't bet on it.

This obsession for alien life is not supported by a single piece of evidence, while the famous question posed by Enrico Fermi—"Where are they all?"—is still waiting for an answer. I believe that the continuous discussion about planets similar to Earth where life could exist or where, in any event, it could easily take root, is a kind of false reassurance for the disasters that we are creating. An assurance that our future, however things go, even if we exhaust the resources of this planet, will be able, somewhere or other, to go on.

Although there exists no evidence of life beyond Earth, try talking about it with anyone who takes an interest in the question and they start trotting out calculations. Beginning with the millions of billions of galaxies in the universe and moving on to the number of likely habitable planets—not counting those with temperatures incompatible with life, those that are too young, or too old, the ones that might seem unfriendly, etc.—they arrive at an extremely high number of planets that host, not *simple* forms of life, but intelligent civilizations, at least as evolved as our own. The mother of all these equations, just to give you an idea of how this argument works, is the famous equation formulated in the 1960s by the astronomer Frank Drake: $N=R \times fp \times ne \times fl \times fi \times fc \times L$.

This equation asserts that the number of civilizations (N) in our galaxy that we might succeed in contacting can be determined by multiplying the average rate of stellar formation in our galaxy (R), the fraction of those stars that have planets (fp), the number of planets that can effectively support life (ne), the number of planets on which life has effectively developed (*fl*), the fraction of planets that have developed intelligent life (*fi*), the number of civilizations that would develop transmission technologies (*fc*), and finally, the estimated duration of these evolved civilizations (*L*). Obviously, depending on the values attributed to the various parameters, it is possible to obtain galaxies pullulating with intelligent life, or, on the contrary, probabilities approximating zero that such life exists.[3]

So let's put the calculations aside. In recent decades, our knowledge about our neighbors in space has grown exponentially. Nevertheless, there has not been any evidence of the existence of life. In the summer of 2015, NASA's space probe *New Horizons* made it to within 7,800 miles of Pluto, the most distant of our planets,[4] sending back to us, as the crowning achievement of a long series of explorations, the first direct information and up-close photographs of this distant planetary neighbor of ours. Another probe has landed on the comet 67P/Churyumove-Gerasimenko. The space probe *Juno* has entered into orbit around Jupiter. For years, the two rovers *Opportunity* and *Curiosity* have been transmitting data on the composition of the soil on Mars, and they have been joined, very recently, by a third vehicle, *Insight*, that will study the subsoil of Mars.

The most interesting result, in my view, of this incessant exploration of our solar system is that the composition of each of the places visited appears to be much simpler than that of Earth. The complexity of our planet comes from life. Living beings are so interwoven with the fabric of the Earth that, outside of some apocalyptic science fiction novel, trying to imagine it as sterile is simply impossible. If it were deprived of life, Earth would resemble something halfway between Venus and Mars. Would it still be blue? Probably not. Certainly, it would not be green. What would be the effect on the planet of the absence of free oxygen? The oxygen that we breathe, in fact, is produced entirely by living beings. To be more precise, by those able to perform photosynthesis. What effect would the absence of oxygen have on the water, the rocks, the soil of our planet? Nobody is able to answer this question.

The truth is that much of what we see on Earth is the product of the actions of living organisms. Rivers, seacoasts, even mountains are designed by the actions of life. The white cliffs of Dover are formed, just as so many other coastal bluffs, from the sedimentary accumulation of the skeletons of countless coccolithophores (unicellular algae coated with scales of calcium carbonate). Much if not all travertine marble is formed through the action of algae. Pyrite and marcasite in sedimentary rocks derive from the reduction of bacterial sulfate. In short, calling our planet Gaia and considering it a single living being is not at all a naive theory, as it has been perceived by so many in the past, but rather a serious way of interpreting the importance and the function of life for Earth.

In 2013, on the basis of solid scientific findings, Bob Holmes described in the magazine *New Scientist* a possible scenario for the future of Earth in the event life were to become extinct.[5] Without plants and other photosynthesizing organisms, the production of oxygen would rapidly cease, and increasing amounts of carbon dioxide would accumulate in the atmosphere. Rising temperatures would melt the polar ice caps. Soil would slide into the oceans and seas for lack of structure, leaving a surface of bare rock and sand very similar to the photos of the surface of Mars sent back to us by the rovers. Over the span of several dozen millions of years, Holmes hypothesizes a planet subjected to an uncontrolled greenhouse effect with conditions similar to those on Venus, so extreme as to make the Earth permanently uninhabitable.

Enough, but that brings me back again to Fermi's question: "Where are they all?" I can't help thinking that the idea that life is so common in the universe is at least partly the consequence of the meager consideration that, deep down, we have for our marvelous planet. Paradoxically, because we live here, we think it must be something rather common.

Do you know the theory of the filter bubble? It's been making the rounds ever since Trump was elected. Were you shocked by the fact that Donald Trump was elected president of the United States? That means that you live in a bubble that prevents you from correctly perceiving reality.

In its original form, the theory of the filter bubble was explained in 2011 by Eli Pariser, in his book *The Filter Bubble: What the Internet Is Hiding from You*. Put simply, Pariser's thesis is that by forming our opinions on the Internet we run the risk of isolating ourselves from other information that is farther away from our cultural and ideological world (our bubble). By using information garnered from our past searches, our contacts, the sites we have visited, the artificial intelligences that administer many of the main Internet sites offer us only what they have reason to believe will interest us. In this way, they isolate us de facto from any exposure to new ideas or opinions outside of our customary worldview and shape our perception of reality. A valid theory, but I would not limit it to the Internet. The truth is that each of us, Internet or not, lives inside a bubble, frequenting people who think as we do, with tastes and attitudes in conformity and compatible with our own. Living inside our bubbles, we believe that what we perceive as normal and shared represents all of reality. And then Trump arrives on the scene to make us realize that it's not true.

So now that we understand what a bubble is, let's expand its coverage to include the whole human community. We all live inside a bubble of life. We humans are living, plants are living, insects, fish, birds, and microbes are living. There is no place on Earth without myriad life-forms. Our bubble is so immersed in life as to make us believe that this is the normal condition of the universe. We are not able to imagine ourselves as the depositories of a unique and fortunate fate. But, in reality, we could very well be inside the fortunate bubble of the beneficiaries of an enormous, incommensurable stroke of luck. The only bubble in the universe formed of living beings. In other words, the only bubble there is.

I know, just saying it seems impossible. It's almost as though someone announced that we had won the first prize of a gazillion dollars in the galactic lottery. Nobody endowed with common sense would believe such a thing. Just like Marie Antoinette, who (legend has it) couldn't understand why the people didn't eat cake. Errors of perception that can cost you your head.

Having reached clarity regarding the immense good fortune of which we are the beneficiaries, the next step is to understand who it belongs to. Who is in charge of this common home? In other words, to whom does its sovereignty pertain? Our most obvious response is that Earth belongs to humans. Or rather, that *Homo sapiens* is the only species entitled to dispose of the planet as befits its needs. This affirmation is so banal that it needs no further evidence to support it. When in the world has the destiny of other species been seen as a limitation on our actions? We have always defined ourselves as the Lords of the Earth, and even if the most progressive among us might feel a bit of shame in considering ourselves the lords of anything, this is nonetheless our most intimate conviction. You'll see.

The Earth is our thing. We have divided its surface into nation-states, and we have assigned its sovereignty to various human groups, who in turn have entrusted it to an extremely limited number of people. It is they, therefore, who hold the real sovereignty over the Earth. *A few people are responsible for the sovereignty of the only planet in the universe on which life exists.* I don't know how much the absurdity of this proposition strikes you, but at times, when I think about it, I go into a dizzy spell and feel as though I've been displaced into one of those infinite parallel universes where logic doesn't work in the way we're used to. A universe governed by crazy laws, even though less fascinating than those of Alice's Wonderland. First of all, what is the source of this investiture that makes us Lords of the Planet? Is it ours by birth or by divine right? Or perhaps by our manifest superiority over the other species, for whose intellectual shortcomings we, as good tutors, are called upon to compensate? Or maybe it's simply a healthy question of democracy and depends on our numbers?

Leaving aside birthright and divine right, about which no logical verification can be exercised, there remain essentially two possibilities. The first: we are the Lords of the Earth because we are the most numerous species. Let's call this the democratic option. The second: we are the Lords of the Earth because we are better than any other living species on the planet. Let's call this the aristocratic option (which I realize also includes, to the delight of my more nostalgic readers, both birthright and divine right).

Let's start with the democratic option, even though I am sure that most of my cultivated readers already know that this cannot be the solution. We are neither the most numerous species nor the largest by biomass. Humanity, with its more than seven and a half billion exemplars, accounts for a quantity of biomass (or living mass) amounting to *one ten millionth* of the entire biomass on the planet. Of the 550 gigatons (one gigaton equals a billion metric tons) of carbonaceous biomass on Earth,[6] animals constitute about 2 gigatons, with insects accounting for about half of that and fish accounting for another 0.7 gigatons. All the rest, which includes mammals, birds, nematodes, and mollusks, comes to 0.3 gigatons. Mushrooms alone have a biomass six times that of animals (12 gigatons). Plants (450 gigatons) represent more than 80 percent of the biomass on Earth, while humans, with our 0.06 gigatons, count for less than 0.01 percent. It is clear that it is not by virtue of our number that we exercise sovereignty over the planet. By number and importance, sovereignty over the Earth should belong to plants.

Having discarded the democratic option for its obvious untenability, the aristocratic option remains alive. From the Greek ἀριστος, *àristos,* "best," and Κράτος, *cràtos,* "power," we humans are Lords of the Earth because we are *better* than any other species that ever existed. I am sure that the aristocratic option seems the much more convincing and robust. Who among us humans is not intimately convinced of being better than any other living species? No kidding. We might be environmentalists, hippies, greens, mystics, materialists, religious, atheists, anarchists, or realists, but about one thing we all agree: we are better than monkeys, cows, apricots, ferns, bacteria, and mold. In this case too the affirmation seems so obvious as to have no need of additional support. We humans *are* better than any other living species, there's no doubt about it. We are better because our big brain allows us to do things that are impossible for anyone else. After all, thanks to our powerful encephalon, haven't we painted the Sistine Chapel, sculpted the Venus de Milo, conceived the theory of relativity, written *The Divine Comedy*, built the pyramids, and philosophized about our existence? What other living being would be able to do something similar? What other species could ever ask itself who has sovereignty over the planet? There can be no doubt about it: humans are better than any other living species!

It is by virtue of this absolute preponderance that we possess Lordship over the Earth. Still, let's try for a moment to shift our gaze away from the splendor of our uniqueness. Once we are no longer bedazzled by our marvelous human conquests, we can try to reason about what exactly it means to be *better* than the rest. The concept of "best" inevitably requires an objective. In a hundred-meter dash, whoever finishes the race in ten seconds is *better* than those who finish in eleven. In a high jump competition, whoever jumps two meters is *better* than someone who jumps one meter ninety. Roger Federer is indisputably *better* than any other tennis player. And what goes for athletes, also goes for artists, musicians, and writers. Dostoevsky is *better* than almost any other novelist.

But in the history of life, what does "better" mean? Or rather, does "better" make any sense at all in the history of the evolution of life? Since there must be an objective for the concept to make sense, what is the objective of life? It looks like one of those terrifying existential questions with no way out, but actually the answer couldn't be simpler: the objective of life is the survival of the species. Darwin tells us that evolution rewards those most apt to survive. The best organism, therefore, is the one most apt to survive.

We have come a long way. Now that we know what the objective is, it should be easy to go on to demonstrate our supposed superiority. Any human, in fact, believes that having a brain as developed as ours is surely an advantage in the struggle for survival. But can we be so sure? Why are we so unassailable in our certainty of our superiority? Could it be that we are falling into another one of those many cognitive distortions, such as the filter bubble, which seem to afflict our glorified brain? There exists, for example, a cognitive disorder, known as the Dunning-Kruger effect,[7] that induces individuals not well versed in a subject to highly overestimate their own competence in that field. (To be sure, it is not as though before Dunning and Kruger came along nobody noticed this. From Socrates on there has been a succession of great minds saying "I know I don't know," but apparently being reminded of it is never superfluous.)

In any case, it is always better to rely on objective data rather than merely declaring ourselves superior, with the risk of falling into the Dunning-Kruger effect. Since we have said that the objective of life is survival, it follows that the species that are supposedly better than the others are those that better succeed in reaching this objective.

All right, by now the question is clear enough: all we have to know is how long a species survives on Earth and, by comparing it to humans, we should be able to establish a ranking of the best species. It is not easy to obtain accurate data on the average life of species, nevertheless reliable estimates tell us that, among animals, survival goes from the ten million years of invertebrates to the one million years of mammals.[8] Obtaining data on the vegetable kingdom is more complicated, because on average plants survive much longer than animals. The *Gingko biloba* is probably more than 250 million years old, the equiseta (horsetails or puzzle grass) were already widespread 350 million years ago. One fern, the *Osmunda cinnamomea*, has been found in fossil rocks from seventy million years ago. In general, it is estimated that the average life of a species, whether animal or vegetable, runs to five million years.

Now that we have our data in hand, let's ask ourselves how much longer we imagine that humans can survive as a species. Obviously, in this case the data really can't help us. However, I am sure that, if we were to ask the same people who are intimately convinced of the superiority of humans whether they believe we will survive for another 100,000 years, the responses would not be all that optimistic. How come? Why do we perceive as unlikely the chances that our species will survive even for just another 100,000 years when in order to reach the average of the other living species we could legitimately wait another 4,700,000 years?

I believe it stems from the disasters that we have managed to create on the planet in a span of time as incredibly brief as the last 10,000 years, that is, since the moment that humans, by inventing agriculture, began to affect profoundly the environment in which we live. We do not believe that we will be able to survive so long as a species because we are aware that our big brain, of which we are so proud, has managed to concoct, in addition to *The Divine Comedy*, also a countless series of dangers that could sweep us off the planet at any moment. Therefore, the monkeys, cows, apricots, ferns, bacteria, and molds that we talked about earlier will keep on becoming extinct only in coincidence with apocalyptical catastrophes, whose frequency on Earth is measured in millions of years, while we risk disappearing at any moment. If we were to vanish tomorrow, what would remain in a thousand years or a hundred thousand, or in millions of years, of the Sistine Chapel, the Venus de Milo, the theory of relativity, *The Divine Comedy*, the pyramids, and all of our philosophical ideas? Nothing. Who would care about all of these wonders?

That is why the very wise Nation of Plants, born hundreds of millions of years before any human nation, guarantees to all living beings sovereignty over Earth. The Nation of Plants wants to ensure that the members of a very presumptuous single species cannot extinguish themselves ahead of time, thus demonstrating that their big brain was not an advantage after all but an evolutionary disadvantage.

• ARTICLE 2 •

THE NATION OF PLANTS SHALL RECOGNIZE AND PROTECT THE INVIOLABLE RIGHTS OF NATURAL COMMUNITIES AS SOCIETIES BASED ON THE RELATIONSHIPS AMONG THE ORGANISMS THAT COMPOSE THEM.

I am sure that many of the erudite readers of this little book know *On the Origin of Species* by Charles Darwin inside and out. If there is someone who still has this gap in their education, you are urged to fill it without any further delay. Darwin's book is fundamental for understanding how life works. And it is surprising to think how this book, which literally changed the history of the world, is actually only a summary of the countless observations that Darwin gathered for decades throughout the scientific disciplines and throughout the world in support of his theory of the evolution of living species. His plan, in fact, was to write a colossal and minutely detailed work that was meant to report all the fruits of his decades of research. It would be a work invulnerable to any and all criticism.

As is well known, things did not work out that way. Alfred Russel Wallace's announcement that he had arrived at Darwin's same conclusions regarding evolution induced Darwin to change his plans and summarize in *Origin* his most brilliant and most evidentially supported deductions, leaving the rest of the material for subsequent elaboration. Nevertheless, the enormous corpus that he was working on did not go to waste. On the contrary, the first two chapters of his magnum opus, which was supposed to be entitled simply *Natural Selection*, became the two volumes of *The Variation of Animals and Plants Under Domestication*, and much of the rest of the material was readapted in the elaboration of his later works. In any event, in the third chapter of *On the Origin of Species*, dedicated to the famous "struggle for existence" that is the dominant motif of the whole book, Darwin tells a marvelous story of relationships. This story is essential for understanding both the bonds between living beings and how difficult it is to imagine the consequences of intervening in those relationships.

Darwin writes: what animals could you imagine to be more distant from one another than a cat and a bumblebee? Yet the ties that bind these two animals, though at first glance nonexistent, are on the contrary so strict that were they to be modified, the consequences would be so numerous and profound as to be unimaginable. Mice, argues Darwin, are among the principal enemies of bumblebees. They eat their larvae and destroy their nests. On the other hand, as everyone knows, mice are the favorite prey of cats. One consequence of this is that, in proximity to those villages with the most cats, one finds fewer mice and more bumblebees. So far so clear? Good, let's go on.

Bumblebees are the primary pollinators of many vegetable species, and it is common knowledge that the greater the amount and the quality of pollination the greater the number of seeds produced by the plants. The number and the quality of seeds determines the greater or lesser presence of insects, which, as is well known, are the principal nutriment of numerous bird populations. We could go on like this, adding one group of living species to another, for hours on end: bacteria, fungi, cereals, reptiles, orchids, would succeed one another without pause, one by one, until we ran out of breath, like in those nursery rhymes that connect one event to another without interruption. The ecological relationships that Darwin brings to our attention tell us of a world of bonds much more complex and ungraspable than had ever previously been supposed. Relationships so complex as to connect everything to everything in a single network of the living.

There is a famous story along these lines told for the first time by the German biologists Ernst Haeckel and Carl Vogt. As the story goes, the fortunes of England would seem to depend on cats. By nourishing themselves on mice, cats increase the chances of survival of bumblebees, which, in turn, pollinate shamrocks, which then nourish the beef cows that provide the meat to nourish British sailors, thus permitting the British navy—which, as we all know, is the mainstay of the empire—to develop all of its power. T. H. Huxley, expanding on the joke, added that the true force of the empire was not cats but the perseverant love of English spinsters for cats, which kept the cat population so high. In any event, underlying the joke is the simple truth that all living species are connected to one another in some way or other by relationships, visible or hidden, and that acting directly on one species, or simply altering its environment, can have totally unexpected consequences. Darwin tells us that trying to imagine the final consequences of any alteration in these relationships would be as "hopeless" as throwing up a handful of sawdust on a windy day and trying to predict where each particle would land.[9] History is full of such attempts, almost always gone wrong, to modify the presence or the activities of single species.

Let's take as an example the affair of the color red. When Cortés and his conquistadores first entered the Aztec capital of Tenochtitlán (present-day Mexico City), they found a very rich and very populous city (in Europe at the time only Naples, Paris, and Constantinople had larger populations). In the enormous market square, a quantity of goods never seen before, many of them of great value, were just waiting to be exported to European markets. Among them were bales of finely woven cotton and delicate yarns of an amazing carmine red. The dye used by the Aztecs to produce this incredible tone of red was obtained from a tiny insect, the cochineal, that lives on cactus plants (various species belonging to the genus *Opuntia*, the prickly pear). The color was so beautiful and precious that states under Aztec domination were required to furnish annually to the emperor a certain number of sacks full of cochineals as tribute. A fine brilliant carmine dye was, and still is, obtained from the dried bodies of these insects.

The production of this dye remained, for almost two and a half centuries, a monopoly of Spain, which guarded the secret jealously and made it into a widespread and highly profitable commerce in Europe. The Spanish sold their dye to whoever could afford it, but above all to the English, who soon became its most enthusiastic and passionate buyers. Enamored of Spanish carmine, which they used to color their military uniforms (their famous *red coats*), the English found a way to buy it at a high price even during their frequent wars against Spain, in which those very uniforms were used. As Italians say, the heart will not be ruled. That special hue of carmine provided by the Spanish dyes was essential for the British army. Any other red would have made their coats less red, demeaning the glorious nobility of the uniform. After all, what kind of image would they have projected in battle with faded uniforms? Their enemies would have died laughing; and that was no way to win a war.

For the next 250 years, despite the best efforts of the English to free themselves from this commercial yoke, the secret of that prodigious dye remained unknown to all but a select fortunate few of Spanish producers. But no production secret can stay that way forever, and so in the closing years of the eighteenth century, British spies succeeded in spiriting away the tightly kept formula: in order to obtain the longed-for carmine, you needed cochineals, and to get cochineals you had to have prickly pears. With the right information in hand, all that remained was to find the right place to begin production. There was no shortage of candidates; the empire was enormous and spread over all the continents. The choice fell on the fortunate Australia. Prickly pears had never grown there, but its climate was perfect for their rapid growth, so both prickly pears and cochineals were imported.

The results were not long in coming. The cochineals died immediately on arrival in Australia, while the prickly pears, useless at this point, were abandoned to their Australian destiny. A destiny of conquerors. Unlike the cochineals, the prickly pears found the Australian environment ideal for their dispersion. With no natural enemies or obstacles and with lots of birds to disperse their seeds, in just a few years the plant spread throughout a vast territory. Having arrived in Australia from Brazil in 1788, the prickly pear was dispersed over an estimated seventy-three million acres, and its expansion did not stop there. It went on conquering new territories at an astounding rate of 1.2 million acres per year. Thus, large amounts of cultivated land, farms, pasture, and agricultural areas of Queensland and New South Wales were invaded by prickly pears, driving away farmers and impeding any kind of productive activity. The problem soon became very serious, forcing the authorities, starting in the second half of the nineteenth century, to look for possible solutions.

In 1901, the government of New South Wales offered £5,000 to anyone who came up with an idea to block the invasion. In 1907, even though the reward had been doubled, it seemed that no one was able to provide an adequate solution. Naturally, there was no shortage of far-fetched proposals. Many people came forward with stratagems that were, let's say, radical. Among them: increase the number of rabbits as predators of the prickly pear, another interesting story of species introduction gone awry. Or, another gem, evacuate an enormous area of land and use airplanes to spray mustard gas (the gas widely used in World War I) to exterminate the animal population, which was responsible for the dispersal of prickly pear seeds. Fortunately, neither of these proposals was taken into consideration, and for decades the only weapon against the devastating advance of the species was to cut down and burn the plants.

Then, in 1926, a solution was finally found: an Argentine lepidopteran (moth) known as *Cactoblastis cactorum*, a parasite of various species of *Opuntia*. By nourishing themselves on cladodes (as the modified leaves of prickly pears are called) the moth larvae managed to debilitate the prickly peril in many parts of Australia. This stratagem enjoyed an extraordinary and unexpected success. In a short time, except in the cooler parts of Australia, where the moth spread less effectively, the prickly pear menace was eliminated.

So it all worked out? In part. Although the introduction of the *Cactoblastis* in Australia is often cited as a successful operation, so much so that the community of Boonarga, just east of the city of Chincilla in Queensland, even dedicated its Cactoblastis Memorial Hall to the moth. Nature always wants the last word. Over time, populations of prickly pears resistant to the parasite evolved in Australia, and this is a first, though not fatal, complication that will, however, require a more careful control of the cactus population in the future. But the second and more important difficulty is that the Australian success in the use of the lepidopteran induced many other nations with analogous prickly pear problems to go down the same road, with totally unexpected results. As Darwin advised us, trying to predict what will happen in a situation like this is like trying to predict where a piece of sawdust will land on a windy day.

In the 1960s the *Cactoblastis* was introduced to the Caribbean islands of Montserrat and Antigua as a control agent of the local cactus populations. In Australia, the sawdust fell in the right spot, but in Central America, it didn't. The moth, in fact, using all kinds of carriers, spread quickly to Puerto Rico, Barbados, the Cayman Islands, Cuba, Haiti, and the Dominican Republic. Through the importation of prickly pears from the Dominican Republic, it arrived for the first time in Florida in 1989, and from there it began to spread at a velocity of over a hundred miles per year along the coast of the Gulf of Mexico. During its expansion, by now completely out of control, this parasite has endangered many cactus populations in the United States and the Caribbean, threatening entire ecosystems, some of them unique. A classic example is the attack on the prickly pear on the Bahamian island of San Salvador, one of the main sources of food for the only extant populations of *Cyclura* iguanas.

And as if all this were not enough, hurricanes, involuntary transport, and trade have recently transported the *Cactoblastis* to Mexico, where it has been sighted for the first time on the island of Mujeres, just off the Yucatan peninsula. In Mexico, unlike in Australia, the prickly pear is a vital plant. It even appears in the national emblem and on the flag. Its fruit and cladodes are a staple food for the population. Prickly pears are used to feed livestock in periods of drought, and some species of *Opuntia* are still used by the cochineal dye industry. If the *Cactoblastis* were to spread to the Mexico mainland, the damage would be enormous.

But no other natural disaster provoked by humans following rash decisions based on inadequate knowledge of natural relationships will ever be able to rival what Mao Tse-Tung accomplished in the late 1950s. Between 1958 and 1962, the Chinese Communist Party led an economic and social movement in the whole country that came to be known as the Great Leap Forward. This was an enormous collective endeavor meant to transform China in just a few years from an agricultural nation into a great industrial power. The movement's results, unfortunately, fell dramatically short of what had been hoped. The reforms through which the party intended to effect this radical national change involved every area of Chinese life, and some of them had devastating effects for the country.

In 1958, Mao was rightly convinced that some of the scourges that had plagued the Chinese for centuries had to be eradicated immediately and in a radical fashion. Keep in mind that when the Communists took power in the autumn of 1949, they found themselves governing a nation gravely distressed by a soaring incidence of infectious diseases: plague, cholera, measles, tuberculosis, polio, and malaria were endemic in most of the country. Cholera epidemics were very frequent, and the infant mortality rate ran as high as 30 percent.[10]

The creation of a national health service and a massive vaccination campaign against plague and measles were the first, meritorious, actions undertaken to improve the situation. Water purification and sewage treatment infrastructure was installed throughout the country, and imitating what had been done previously in the Soviet Union, health care personnel were trained and sent into rural areas to serve as proper health care administrators, educating the population in basic health and hygiene practices and treating diseases with all available resources. But, obviously this wasn't enough; the diffusion of carriers that spread disease had to be curtailed: mosquitoes, responsible for malaria; rats, spreaders of plague; and, finally, flies had to be exterminated. These three scourges from which China had to be liberated were soon joined by a fourth: sparrows, which by eating fruit and rice cultivated laboriously in the fields were one of the most terrible enemies of the people. Chinese scientists had calculated that each sparrow ate ten pounds of grain per year. So for every million sparrows killed, food for 60,000 people would be saved.

This information was the basis for the "Four Pests Campaign," and sparrows were public enemy number one. Today, any proposal for ecosystem modification as radical as this call to eliminate four species from a territory as vast as China would, obviously, be considered ill-considered. But in 1958, lots of people thought it seemed like a good idea. So the party's campaign to recruit the citizenry to combat these four pests was begun. Millions of posters were printed up illustrating the necessary eradication and the means to implement it.

For the battle against sparrows, the people were told to give no quarter and to use all available means. One of the directives was to frighten the sparrows with noise, produced in any way possible, so they would be forced to fly constantly without ever coming to rest, until they fell to the ground exhausted. Pans, casseroles, gongs, rifles, trumpets, horns, plates, tambourines—any source of noise was put to use. Here is a description of what happened by a Russian observer, Mikhail A. Klochko,[11] who was working as a consultant in Beijing when the four pests campaign was launched:

I was awakened early in the morning by the sound of a woman screaming. Rushing over to the window, I saw a young woman running back and forth on the roof of a nearby building, frenetically shaking a bamboo pole with a large sheet tied to it. Suddenly, the woman stopped yelling, apparently to catch her breath, but an instant later, down at the end of the street, a drum started beating, and the woman went back to her blood-curdling screams and the mad shaking of her peculiar banner. This went on for several minutes. Then the drums stopped beating and the woman fell silent. I then realized that, on all the upper floors of my hotel, women dressed in white were waving sheets and towels that were meant to prevent sparrows from landing on the building. This was the opening of the Great Sparrow campaign. All day long I heard drums, gunshots, and screams and saw fluttering sheets, but never at any time did I see a single sparrow. I cannot say whether the poor birds had perceived the mortal danger and flown off in advance to safer terrain, or if there had never been any sparrows in that place. But the battle went on without abatement until noon, with the entire staff of the hotel mobilized and participating: porters, front office managers, interpreters, chambermaids and all the rest.

Although Klochko's account makes it seem that all this activity was not very effective, the actual results were, unfortunately, devastatingly successful. The government acclaimed the schools, working groups, and governmental agencies that achieved the best results in terms of number of pests killed. The estimates provided by the Chinese government, totally unreliable for their enormity, indicated a billion and a half rats and a billion sparrows killed. Even though they are enormously exaggerated, these figures nevertheless tell us of a massacre whose dramatic consequences would soon be evident. Sparrows, in fact, do not feed exclusively on hulled grains. On the contrary, their main food supply are insects.

In 1959, Mao, realizing his mistake, replaced the sparrows as a target pest with beetles, but the damage had already been done. The almost total lack in China not only of sparrows (which had to be reintroduced from the USSR) but of practically all other birds led to an immeasurable increase in the insect population. The number of locusts began to increase exponentially, and immense swarms of insects making their way through the fields of China destroyed most of the crops. From 1959 to 1961, a series of ill-starred events partially related to natural disasters and partly caused by the mistaken reforms of the Great Leap Forward (the idea to exterminate the sparrows being one of the worst), led to three years of famine so harsh that it caused the deaths of an estimated 20 to 40 million people.

Playing with something whose working mechanisms are not well known is clearly dangerous. The consequences can be completely unpredictable. The strength of ecological communities is one of the engines of life on Earth. At every level, from the microscopic to the macroscopic, it is these communities, understood as relationships among the living, that allow life to persist. Already in 1961,[12] one of the first studies able to rely on the use of electronic calculators for the performance of the numerous and complex calculations required by its models demonstrated that floating communities of microscopic organisms in Virginia's York River were not in fact at the mercy of the environment. On the contrary, together they turned out to be five times more stable than the physical environment itself.

Ecological communities are the basis of life on Earth. The entire planet should be considered as a single living being—this is the Gaia theory—whose balancing mechanisms (the more technical term is homeostasis) are able to generate the forces and counterforces needed to temper the swings in a continually changing environment. In other words, something like the mechanisms that keep our body temperature constant even though the temperature of our surrounding environment is constantly changing. Life has evolved through these communities, and it will continue to exist and evolve only if humans are prohibited from interfering with them. That is why the Nation of Plants recognizes as an inalienable right the inviolability of any natural community.

• ARTICLE 3 •

THE NATION OF PLANTS SHALL NOT RECOGNIZE ANIMAL HIERARCHIES, WHICH ARE FOUNDED ON COMMAND CENTERS AND CENTRALIZED FUNCTIONS, AND SHALL FOSTER DIFFUSE AND DECENTRALIZED VEGETABLE DEMOCRACIES.

Plants and animals separated between 350 and 700 million years ago in a decisive period for the history of the evolution of our planet. In correspondence with this fundamental crossroads, in fact, life would take two divergent roads that would lead, on one hand, to the birth of plants and, on the other, to the birth of animals. The former, thanks to their prodigious capacity for photosynthesis, which makes them energetically autonomous, do not need to move around in search of food. The latter, on the contrary, obliged in order to survive to prey on other living organisms, are bound to movement in a constant search for that same chemical energy that plants originally glean from sunlight. This initial divergence produced very different organisms in terms of organization and functioning.

To be rooted to the ground, with no chance of moving from one's place of birth, has fundamental consequences. Plants do not flee from predators, do not go hunting for food, do not move to more comfortable environments. Plants do not have the possibility to adopt the primary strategy that animals adopt to resolve all kinds of difficulties: movement. But if you can't run away how is it possible to resist predators? The trick lies in not having any fundamental single or double organs, while at the same time distributing throughout the entire body all those functions that animals concentrate in specialized organs. Animals see with their eyes, hear with their ears, breathe with their lungs, think with their brains, and so on. Plants see, hear, breathe, and think with their whole bodies. An essential difference: concentration versus distribution, whose consequences for the life of us animals are not immediately obvious.

Naturally, the extreme fragility of our bodies is evident to us all. A slight malfunction of any one of our organs is enough to put our survival at risk. That is one of the consequences of the way we are organized, though it's not the only one, nor, I believe, the most important. Being built with a brain that presides over the functioning of various specialized organs has influenced practically every kind of organization or structure that humans have invented. In everything we do we replicate this centralized and hierarchical organization. Our societies are constructed according to this same scheme. Our companies, offices, schools, armies, associations, parties—everything—are organized according to pyramidal structures. Our very tools, even the most modern ones like the computer, are simple synthetic analogues of ourselves: a processor that mimics the functions of our brain, that governs circuit boards (hardware) that imitate the functions of our bodily organs.

The only advantage to this type of organization is speed. A chief, who is the only one entitled to decide, should be able to rapidly determine the actions to be carried out. This quality of centralized organizations, while it may ensure to animals' bodies the necessary speed of action, nevertheless fails badly in human practice. Every hierarchical organization, in fact, evolves its own bureaucracy, or rather a group of people whose job it is to transform into routine the mechanism of transmitting commands through the various levels of the hierarchy. Transmission from one level to another, beyond being inevitably subject to error, requires time, thus eliminating speed of action, that is, the only real advantage ascribable to a centralized organization. Its countless disadvantages, however, remain intact, from the fragility of the organization, which is subject to collapse upon removal of any of its essential organs, to the distance between the decision-making center and the places where the decisions are implemented. That is certainly not the end of it. Problems inherent to the existence of bureaucracy, the fundamental connective fabric of any hierarchical organization, are numerous, and one worse than the other. Taking account of them might help us to understand just what kind of a fix we are in.

The problems we face are inevitably tied to the very existence of the hierarchical chain. Take for example, the Peter Principle, which you have probably heard talked about and probably imagined was really a joke, a gimmick to describe in humorous terms the typical situation that comes to be created in the worst public bureaucracies. Actually, however, it describes a serious problem present in all bureaucracies. Conceived by

Lawrence J. Peter in 1969,[13] this principle observes that the people in a hierarchy tend to rise to their "level of incompetence."

What does this mean? Imagine a perfect hierarchical organization, in which each member of the organization is promoted from one level to the next higher level only by virtue of merit. A utopian organization, where jealousy, politics, rancor, friendship, family, demographics, or personal relationships have no influence at all on the way people are promoted from one level to the next. Let's abstract ourselves for the moment from our squalid world of careerist conniving, personal envy, and resentment, and free our minds all the way to the empyrean heights of this miraculous organization, in which merit, and merit alone, is the driving force of its members' careers.

It would appear to be the perfect organization, am I right? And yet, by the sole fact of its being hierarchical, Peter tells us, an organization of this type is incapable of functioning. In fact, any member of the pyramid who was competent at a certain level of the organization would, simply by virtue of having achieved that level of competence, be promoted to a higher position in the hierarchy, one requiring different skills and abilities. Should the person promoted not have the skills and abilities required by the new level, he or she would remain at that level (known as Peter's Plateau). Alternatively, by showing themselves competent at the newly reached level, they would be promoted again, until reaching, by the very nature of things, a level where they were not competent, and their career would be blocked. In either case, the inevitable result could not be other than that expressed by the Peter Principle: *in a hierarchy every employee tends to rise to his or her level of incompetence.*

This principle was already intuited a century earlier by the Spanish philosopher José Ortega y Gasset, who wrote, "all public employees should be demoted to their immediately lower level, as they have been promoted until turning incompetent." Although the book in which Peter expounded this principle for the first time was written with a satirical intent, the conclusions it reaches are anything but whimsical, as has been confirmed by a long series of studies conducted in the years since. One of the most recent, published in 2018, examined the dossiers used for the promotion of employees in 214 American companies, discovering that the companies tended to promote to management people who in their previous positions had demonstrated strong abilities in sales but who had no or only negligible management skills.[14]

The Peter Principle is certainly not the only problem connected to bureaucracies and, indirectly, to all other hierarchical organizations. Bureaucracies are created, in fact, to respond to the need to channel orders to the various levels of the organization. But, once created, a bureaucracy tends to grow out of control, multiplying its members as long as the possibility exists, or rather as long as there are resources still to be consumed. In 1955, C. Northcote Parkinson, in a celebrated essay first published in *The Economist* and later as a book,[15] stated what would later become known as Parkinson's Law. Formulated on the basis of the behavior of gases, which expand as long as there is available volume, Parkinson's Law affirms that bureaucracy expands as long as it is possible to do so. In support of his law, Parkinson cites a series of examples and very persuasive empirical data. Among these is the continuous and steady increase in the number of employees working in the Colonial Office of the British Empire—notwithstanding the decline over time in the number of colonies—which reached its peak when, there being no more colonies to administer, it was absorbed by the Foreign Ministry. According to Parkinson's Law, this happens inevitably in every bureaucracy, whether or not the amount of work to be done remains the same, diminishes, or even disappears. The reason is to be found in the simple fact that the members of a bureaucracy tend to multiply their subordinates rather than their possible rivals.

Let's try to clarify this point. A worker who has a certain amount of work to do and who realizes that he or she will not be able to complete it, because the amount of work has increased or simply because he or she no longer feels like doing it, has three possible strategies to resolve the problem: 1) resign, 2) decide to share the work with a colleague, 3) hire two more subordinates (they must necessarily be two because having only one would turn that subordinate into a rival, and we would be back to a case 2 situation). Now, let's quickly analyze the consequences of each of these options. The first is easily discarded in that it would leave the worker without a job. The second would lead to the creation of a potential rival for an eventual promotion, while the third is the only strategy that would allow the worker to maintain his or her position and career opportunities, while working less. Inevitably, after a little while, the two newly hired workers will find themselves in the same situation, and the only possible solution will be to hire two subordinates for each of them. Thus it is easy to see that by following these diabolical dynamics, sooner rather than later, seven people will be doing the same amount of work that was previously done by only one.

Parkinson's Law can also be expressed in mathematical form, by way of a simple little formula whose resolution tells us that according to Parkinson the annual increase in the members of an organization will be, invariably, between 5.17 and 6.56 percent. And it is amazing to see that many bureaucratic organizations *really do* grow at rates very close to those predicted by Parkinson's Law. In short, bureaucracy is one of the worst consequences of animal (that is, centralized) pyramidal organizations with a chain of command. In the end, as the sociologist Max Weber writes, every bureaucracy stops serving the society that created it, becoming an end in itself, growing like a foreign body, implementing measures that protect itself and imposing dysfunctional rules that serve only to justify its own size.[16] The damage done by the hierarchical bureaucracies inspired by animal architecture would be enough in and of itself to make Article 3 of the Constitution of the Nation of Plants a shining example of wisdom.

Unfortunately, bureaucracies are only one, and not at all the worst, of the many problems afflicting hierarchical and centralized organizations. We will be looking at some of the others in the following pages. One of the lesser-known problems of hierarchical organizations is that they are bad for your health. In 1967, a study was begun in Great Britain on the physical and mental health of British public employees. The so-called Whitehall study focused on public employees as representative of a middle class in good health and not exposed to the direct health and safety risks to which other categories of workers, such as soldiers or miners, might be exposed. The British Civil Service, like most large-scale organizations, is very hierarchical. Government employees are ranked from one to eight according to their level in the hierarchy, and the amount of their compensation is directly related to their rank: the higher their rank, the higher their pay and benefits. Initially, the study examined for a period of ten years more than 18,000 male employees between the ages of 20 and 64. A second study examined another 10,000 public employees from 35 to 55, two-thirds men and one-third women.[17]

The main result of this series of studies demonstrated incontrovertibly that there was a direct correlation between the level reached by the worker and the mortality rate: the lower the ranking, the greater the mortality rate. Employees at the lowest level of the hierarchy (messengers, guards, and so on) had a mortality rate *three times higher* than that of employees at the highest level (managers). This effect, which has since then been verified in many other analogous studies, has been named "status syndrome."[18] Furthermore, these studies revealed that the ranking achieved in the bureaucracy was indirectly correlated to a series of pathologies, including some forms of cancer, heart disease, gastrointestinal disorders, depression, backache, and so on. Now, certainly, many of these pathologies were associated with risk factors such as obesity, smoking, high blood pressure, and lack of physical exercise, directly related to social class and, therefore, to low income, without connections of any kind to one's rank in the hierarchy. But what remained unexplained was that these risk factors contributed only in part to the final results. Even when these nonhierarchical factors were controlled for, there still remained a risk of cardiovascular disease for employees in the lowest levels of the hierarchy that was 2.1 times higher compared to employees at the highest levels.

The factor that modified the mortality rate so significantly was the substantially higher stress level experienced at lower levels of the hierarchy. This higher stress level in the lower ranks of the hierarchy is directly tied to hierarchical organization, and we share this risk with animals similar to us, such as baboons, who form highly hierarchical groups. Indeed, in alpha males, that is, in monkeys at higher levels in the hierarchy, the blood levels of glucocorticoids (a class of steroid hormones, including cortisol, also known as the stress hormone) were significantly lower than those in the lower rankings of the hierarchy. It was even found that baboons at lower levels of the hierarchy stored fat mostly in their bellies, while alpha males distributed fat uniformly throughout the body. In other words, subordinate monkeys took on a rounder and more passive aspect consonant with their hierarchical ranking, in contrast to their thinner and more muscular superiors.

Then one day it happened that the alpha males in this group of baboons, and many of the other males higher up in the hierarchy died of tuberculosis, leaving the group at half size and made up disproportionately of females and low-level males. For a number of reasons, the group learned a new system of interaction without hierarchies and started teaching this new system to new males who joined the group. From then on, the blood levels of glucocorticoids for members of the group leveled off, showing a significant reduction in stress levels.

So, to recapitulate, hierarchies are also bad for health. Is that all? Not by a long shot. Unfortunately, the worst is yet to come.

All centralized and hierarchical organizations, for example, are inherently *fragile.* Hernán Cortés and Francisco Pizarro, accompanied by a few hundred men, brought down two millenarian civilizations, the Aztecs and the Incas, simply by capturing their leaders: Montezuma and Atahualpa.

Two highly evolved societies with advanced knowledge in numerous fields of science, constituted of millions of people—the city of Tenochtitlán alone, when Cortés arrived there on November 8, 1519, numbered around 250,000 inhabitants—dissolved in the blink of an eye under the assault of the conquistadores.

Obviously, multiple causes concurred in the fall of these two empires. Among them, though rarely cited, was the extreme centralization of power, concentrated in the hands of a few. The Apache, much less advanced than the Aztecs and Incas but endowed with a distributed organizational structure and no centralized power, resisted the Spanish onslaught for centuries, blocking their expansion into the northern part of the continent. But even fragility is not the worst problem of hierarchical organizations.

In 1963, Hannah Arendt published *Eichmann in Jerusalem: A Report on the Banality of Evil,* one of the essential books for understanding the history of the twentieth century. The book is the product of her work as a reporter during the criminal trial of the Nazi Adolph Eichmann, who as the chief organizer of the Holocaust was responsible for the deaths of millions of Jews. Eichmann's defense was centered on obedience to authority. From the court hearings, Arendt draws her conviction that Eichmann, and the majority of his fellow Germans, co-responsible for the Shoah, were not driven by some special disposition to evil, but by their belonging to a hierarchical organization in which the bureaucrats charged with transmitting orders were unaware of the ultimate significance of their actions. At the time, Arendt's assertions seemed unreasonable. The thesis that a hierarchical organization characterized by 1) sufficient distance between one's own actions and the results of those actions, 2) strong authority, and 3) depersonalized internal hierarchical relationships, could re-create the horror of the Shoah seemed totally unacceptable to most observers. What Arendt wrote scandalized the world. Not only could the Shoah happen again, but anyone was capable of being responsible for it. It was a disturbing hypothesis, which only after some time began to be elaborated correctly, but which in the beginning stimulated in a large number of readers a reaction of complete refusal. It couldn't be true that an enormity such as the Shoah was dependent in the first instance on a form of organization. The reactions to Arendt's thesis were violent, and her idea that evil could arise "banally" anywhere and everywhere was rejected flat out.

In the same year that *Eichmann in Jerusalem* was first published, a Yale psychologist, Stanley Milgram, obtained a stupefying series of experimental results, published first in a specialized journal[19] and then ten years later in a book entitled *Obedience to Authority,*[20] which should be read together with *Eichmann in Jerusalem.*

The experiment designed by Milgram was based on the interaction of three people: a scientist, who is the authority figure, a teacher, who carries out the orders of the authority figure, and a student, subject to the decisions of the teacher. Teacher and student are in two different rooms; the student is connected to some electrodes through which the teacher can administer electric shocks. The teacher's task is to instruct the student to repeat some pairs of words. When the student makes a mistake, the teacher punishes the student with electric shocks of gradually increasing intensity, from a minimum of 15 volts to a potentially fatal maximum of 450 volts. Both the student and the scientist are actors, and all the machines are fake.

The real subject of the test is the teacher. What interests Milgram is finding out the number of those disposed to follow the indications given by the authority figure (the scientist) up to the point of punishing the student with potentially lethal shocks.

Without going into the details of the experiment, which in any case can be easily found on the Internet, the results were striking: more than 65 percent of the teachers administered the maximum shock. In a series of variations on the experiment, in which the student was in the same room with the teacher (proximity), or in which two scientists discussed what to do (authority figure weakened), the percentage decreased to less than 20 percent. This was the experimental demonstration, the scientific proof of what had been claimed by Arendt. Although it too was strongly contested at first, Milgram's experiment was repeated in subsequent years in the most varied contexts, always yielding very similar results.

Despite their various negative, or at the very least problematic, aspects highlighted so far, hierarchical organizations, with their perfect reproduction of the architecture and functionality of the animal body, are everywhere. Is it possible that we cannot manage to come up with something different, as would be the case, for example, of diffuse organizations constructed like the body of a plant?

Though they are rare, there are a few important examples of such differently structured human organizations. What's more, they are almost always innovative. The Internet itself, the very symbol of the contemporary world, is constructed like a plant: completely decentralized, diffuse, composed of an enormous number of repeated identical nodes, with no specialized organs.

Try comparing the topography of a root system with any map of the Internet, and the architectural resemblance is unmistakable. Plants, including their root systems, are constructed in modular fashion. Single modules that repeat endlessly to form ever more extensive and complex structures, but without any fundamental center. A root system is built out of an astronomical number of root apices—a tree can have hundreds of billions—which, extending themselves in the soil and exploring for the nutrients and water needed by the plant, form such a complex network as to rival the structural complexity of our own neural networks. Unlike our brain, however, which is incredibly fragile and whose different cerebral areas are specialized for the performance of specific functions, a root system spreads its functions throughout. Therefore, root systems, not having areas specialized in fundamental functions, can easily survive extensive damage to the majority of the components of the network.

Even the architecture of a tree's foliage follows the same rules of diffusion and repetition of similar modules, even though the foliage of each species is so different from the others as to be distinguishable to the naked eye even at great distances. In 1972, Roelof Oldeman, while paddling a pirogue on the Yaroupi River in French Guyana,[21] realized that the trees were made of *reiterated* modules, which represented their precise architectural characteristics. Anyone who observes a sucker or a shoot (that is, one of those very vigorous sprouts produced from latent buds, generally located at the base of a tree trunk) can notice that each of them gathers within itself the general characteristics of the tree. Wherever you look, from the roots to the foliage, you notice that plants are constructed on a diffuse model, the opposite of the centralized animal model. It is an organization that allows for freedom and strength at the same time. In recent years, forms of decentralized organization have been spreading quickly.[22] They provide for diffuse decision-making, where consensus and authority derive from one's own capacity to influence, rather than being conferred from above. In these diffuse organizational models, with no single command center, decisional centers, as in plants, are dispersed and are created spontaneously on the periphery, that is, where they must be in order to resolve problems with precision: where the needs are clearest and useful information is more readily available.

The Nation of Plants, by utilizing only diffuse, decentralized, and reiterated organizational models, has liberated itself forever from the problems of fragility, bureaucracy, distance, sclerosis, and inefficiency typical of the hierarchical and centralized organizations of the animal world.

• ARTICLE 4 •

THE NATION OF PLANTS SHALL UNIVERSALLY RESPECT THE RIGHTS OF THE CURRENTLY LIVING AND THOSE OF FUTURE GENERATIONS.

Whether Burundian, Italian, or Icelandic, humans are the most accomplished predators. Like a lion observing, sleepy and satisfied, the piece of the savannah that is his territory, with the serene awareness that no other animal can contest his sovereignty over it, the human race considers the entire planet as something under its exclusive jurisdiction. Earth, the home of life, the only place we know of in the universe able to host it, is considered by humans as neither more nor less than a simple resource: to be eaten, to be consumed. Something similar to a gazelle in the eyes of an always-hungry lion.

That this resource might come to an end, putting at risk the very existence of our species, does not seem to interest us. Have you ever seen that science fiction film in which some really wicked alien species, after having consumed the resources of countless other planets, swoops down on the Earth like a swarm of grasshoppers from space intent on turning it into a wasteland? Those aliens are us. Only the other planets still left to be destroyed after Earth do not exist. We would do well to understand this as soon as possible.

The consumption of organic material produced by other living beings is typical of animal life. Not being able, as plants are, to capture the energy of sunlight autonomously, animals must perforce rely on the predation of other living beings to ensure their survival. This is why plants are always pictured at the bottom of those typically pyramidal illustrations that we see everywhere bearing the name of the food pyramid, or the ecological pyramid, or the trophic pyramid. Whatever the name, the concept is always the same. There is a pyramid with plants, the producers, occupying the lowest level, and then proceeding upward through the various trophic levels. First, the herbivores that eat plants, then above them the carnivores that eat meat, and then the omnivores that eat both plants and meat, and so on, until you get to the apex predators, which are at the top of the food chain.

I have always found these representations of plants as the lowest level of a pyramid to be rather ungenerous, not to say wrong. It would seem to me more correct that the top should be reserved to the organisms that *produce* chemical energy, rather than those that consume it. I mean, in an automobile isn't the most important part the engine? All the rest is not essential. Well, plants are the engine of life, the essential part; all the rest is just auto body.

Every time that the energy produced by plants is transferred from a lower level to the next higher level of the pyramid (e.g., when the herbivores eat plants) only 10 to 12 percent of the energy is used to constitute new body mass, thus becoming stored energy, while the rest is lost in various metabolic processes. Therefore, at each successive level we will find 10 percent of the energy present at the preceding level. This is a precipitous drop. Just think, if we attribute to the primary producers (plants) an arbitrary energy level of 100,000, the successive levels will be 10,000, 1,000, 100, 10, 1, and so on. In practice, the organisms positioned at the top of the pyramid, the so-called apex predators, are the least *sustainable* in terms of energy that one can imagine.

Students of ecology have been debating for years whether or not humans, on the basis of their diet, should be considered apex predators. Some claim that the inhabitants of the various nations of the Earth have different trophic levels, varying from the 2.04 of Burundi, which has a diet that is almost exclusively vegetarian, and, therefore very close to the level 2 of pure herbivores, all the way to the 2.57 of Icelanders, who, on the other end of the scale, have a diet that is just 50 percent vegetables. For anyone who is interested, these trophic levels would make us comparable to pigs.[23] Other ecologists, instead, believe that humans should be considered the apex predators of every trophic chain.[24]

I have always found this debate fascinating in its futility. It is obvious that humans are the only true apex predator on the planet. What's more, humans' peculiarities make them incredibly more dangerous for other species than any other living being.[25] It is precisely in their activities as apex predators, that is, as the maximum expression of animal life, that humans are consuming non-regenerable resources at an ever-faster pace, and with the waste products of this senseless activity are increasingly polluting air, soil, and water. How dangerous this predatory activity is and how much damage it has already done is hardly even noticed. Sure, there is a lot of talk about global warming, climate change, urban pollution, decreasing biodiversity, and so on, but I do not believe that the gravity of the situation is clear to most people. At least I hope that is the case. If not, it would mean that humanity has lost all sense of its own future.

Many of you will have heard talk about the Anthropocene—I have also written about it recently.[26] The Anthropocene is the name for the geological era we live in, whose dominant feature is the seismic effect of human activity. Humans, for example, through their continuous and uncontrollable need to consume, are so profoundly affecting the characteristics of the planet as to have become the cause of one of the most terrible mass extinctions of all time. In the history of our planet, catastrophes of similar dimensions have happened only after apocalyptic events such as asteroids, volcanic eruptions, inversions of the Earth's magnetic field, supernovas, the rising or lowering of ocean levels, and glaciations. The frequency of such events has been estimated to range from once in every 30 million[27] to once in every 62 million years,[28] and their causes have been hypothesized as depending on circumstances such as oscillations of the galactic plain or the Earth's passage between the spiral arms of the Milky Way.[29]

Throughout its history, Earth has suffered five mass extinctions and a certain number of minor extinctions. The five big ones, identified by Sepkoski and Raup in a noted work from 1982,[30] are: 1) the Ordovician-Silurian extinction, from 450 to 440 million years ago, when two events occurred which eliminated between 60 and 70 percent of all species, representing the second biggest of the five major extinctions in terms of the percentage of genera extinguished; 2) the late Devonian extinction, lasting perhaps some twenty million years, during which about 70 percent of extant species disappeared; 3) the extinction during the transition from Permian to Triassic, 252 million years ago, the most dramatic extinction event to strike the Earth, where from 90 to 96 percent of extant species were swept away; 4) the Triassic-Jurassic transition extinction, 201 million years ago, which eliminated from 70 to 75 percent of species, and, finally 5) the extinction during the transition from Cretaceous to Paleogene (the one in which dinosaurs became extinct), 66 million years ago, which killed off 75 percent of living species.

Today we are in the thick of the sixth mass extinction, an event of such proportions that perceiving its consequences is difficult indeed. The current extinction rate of species is unimaginable. In 2014, a research group coordinated by Stuart Pimm of Duke University estimated the Earth's normal extinction rate, prior to the appearance of humanity, as 0.1 extinct species per million species per year (0.1 E/MSY). The current rate appears to be 1,000 times greater, while models for the near future would indicate an extinction rate as high as 10,000 times normal.[31]

These are the numbers of an apocalypse. Never in the history of our planet, not even during the most catastrophic mass extinctions, has there been such a high rate of extinction and, above all, compressed within such an imperceptible span of time. The past mass extinctions that we know about, as rapid as they were, have always come about over a span of *millions* of years. Human activity, on the contrary, is concentrating its lethal effect on other living species in a handful of years. The entire history of *Homo sapiens* began just 300,000 years ago, less than the blink of an eye compared to the 3.8 billion years of life. Those who worry about the invasiveness of such magnificent vegetable species as the ailanthus, the robinia (black locust), the pennisetum (fountain grass) and others, because of their capacity to drive out native species from their home territories, should be aware that, compared to the invasiveness of *Homo sapiens*, the dangerousness of any other species, animal or vegetable, is nothing more than a joke.

At the end of 2017, 15,364 scientists from 184 countries signed a declaration, "World Scientists' Warning to Humanity: A Second Notice," in which they affirmed that "we have unleashed a mass extinction event, the sixth in roughly 540 million years, wherein many current life forms could be annihilated or at least committed to extinction by the end of this century."[32] We might be tempted to just shrug our shoulders at this news. Many, deep in their hearts, might think, "We have destroyed entire human civilizations, why worry about the disappearance of a number, high as it might be, of animal and vegetable species? We will survive without any problem."

I think this is actually the biggest danger: thinking that what we are doing does not directly affect the preservation of our civilization, not to mention the survival of our species. How could the extinction of plants, insects, algae, birds, and various mammals have an impact on our survival? OK, it's sad that rhinos, gorillas, whales, elephants, bananas, monk seals, lightning bugs, and violets are becoming extinct, but, after all, how many of us have ever even seen one? For us city dwellers, nature is the stuff of documentaries, it has nothing to do with us. What interests us is the interest-rate spread, the GDP, Euribor, the Nasdaq—these are the things that could cause the collapse of civilization as we know it.

Wrong! I repeat, it is the very idea, so widespread as to have become a cliché—that we humans are outside of nature—that is truly dangerous. The extinction of such a high number of species, in such a short time, along with ongoing and accelerating declines in species' populations, is something whose consequences we cannot estimate. Rodolfo Dirzo, a professor at Stanford and an expert on species interaction, writes, "Our data indicate that beyond global species extinctions Earth is experiencing a huge episode of population declines and extirpations, which will have negative cascading consequences on ecosystem functioning and services vital to sustaining civilization. We describe this as a 'biological annihilation' to highlight the current magnitude of Earth's ongoing sixth major extinction event."[33]

Now, it's true that Cassandras have never been well liked by anybody, and often we even forget that Cassandra—the original—the unheeded prophetess, was right! Being aware of the disaster that our consumption is creating should make us all more careful about our individual behavior, but also angry toward a model of development that in order to reward a few is destroying our common home.

• ARTICLE 5 •

THE NATION OF PLANTS
SHALL GUARANTEE
THE RIGHT TO
CLEAN WATER, SOIL,
AND ATMOSPHERE.

At the start of the twentieth century, a famous Russian botanist, Kliment Timiryazev, wrote in his book *The Life of the Plant* that plants should be considered the connecting link between the sun and the Earth. Without plants, in fact, the sun's energy would not be transformed into the chemical energy that feeds life. That's not all. Plants perform a fundamental and continuous labor of depollution by absorbing and degrading many of the contaminating compounds produced by humans. Unlike us humans, who in carrying out all of our normal activities inevitably pollute the soil, water, and atmosphere of the planet that hosts us. But let's go back and take a look at this issue from the beginning and try, if possible, to clarify exactly what the problem is.

Every living being needs to obtain from some energy source the amount of energy it needs to survive. The energy present on planet Earth comes from three principal sources: the sun; the primordial heat produced by the original formation of our planet; and, finally, the heat produced by the radioactive decay of some of the materials composing the core and crust of the Earth. For the purposes of our discussion, we can easily forget about the contribution of geothermal energy and concentrate, instead, on the energy of the sun, the real font of energy for life on Earth. Even the energy that we obtain from the combustion of coal or oil, in fact, is nothing more than solar energy originally captured and fixed by plants (understood here in the general sense of photosynthetic organisms). Similarly, the energy generated by the wind and by ocean currents or waves is still, at its origin, derived from solar energy. In sum, hoping for the indulgence of physicists and geologists, we can approximate that, for our purposes, practically all the energy on the planet comes from the sun.

Now that we have simplified the problem to its principal terms, we can return to plants, to the central role they play in ensuring the survival of organisms and to what Timiryazev stated in his book. To be precise, he identified not plants but a specific organelle found in green cells—the chloroplast—as the real connecting link between the Earth and the sun. Timiryazev argued that without the chloroplast—inside of which the miracle of photosynthesis takes place—there would be no conversion of solar energy into sugars (chemical energy). Perhaps it would still have been possible to find some minimal form of life on Earth, but it could hardly have been a life as complex or as enormously plentiful as we are used to.

Through photosynthesis and thanks to the energy of the sun, plants fix the carbon dioxide present in the atmosphere, making sugars, that is, highly energetic molecules, and producing oxygen as a waste product. The average annual amount of energy produced by photosynthesis on a planetary level is around 130 terawatts,[34] or about six times greater than the total current energy consumption of human civilization.[35] Talking about the cycle of carbon in his *The Periodic Table,* Primo Levi writes: "If the elaboration of carbon were not a common daily occurrence, on the scale of billions of tons a week, wherever the green of a leaf appears, it would by full right deserve to be called a miracle."[36]

Thanks to this miraculous process, life has been able to spread and prosper. Practically speaking, photosynthesis is solely responsible for the entire production of organic matter created through biochemistry, so-called primary production. The amount of material generated by plants is hard to imagine. The most careful estimates indicate a primary production for the Earth of 104.9 petagrams of carbon per year (PgC yr^{-1}). Or, for those who do not recall right off that a petagram amounts to a number of grams equal to a 1 followed by 15 zeros, 104.9 billion metric tons of carbon are fixed by plants every year. Of these, 53.8 percent are produced by terrestrial organisms, while the remaining 46.2 percent are produced by plant life in the oceans.[37] This enormous quantity of organic substance, generated by way of photosynthesis, is the engine of life on Earth.

Once it is produced by plants, this chemical energy—you can, if you will, imagine it in the form of food, or coal, or petroleum—is used as fuel by the rest of animal life, which uses it in the amount necessary to assure its survival. It is also used by humans, who, however, use it in amounts greatly exceeding survival levels as the source of energy for their own development. When this carbon burns, it inevitably produces residues that alter the equilibrium of the environment and pollute it. Carbon dioxide, for example, is produced every time combustion happens. Whether we burn sugars or fats to obtain the energy essential for our bodies, or we burn petroleum, gas, coal, wood, or any other kind of combustible fuel originally produced through photosynthesis, the end result is the production of carbon dioxide.

Human activities emit about 29 billion tons of carbon dioxide per year, while, by comparison, volcanoes emit a hundred times less, between 200 and 300 million tons. The carbon dioxide that accumulates in the atmosphere is the primary cause of the greenhouse effect, and therefore of global warming. Through their activities—especially the combustion of fossil fuels and deforestation—humans, since the beginning of the Industrial Revolution, have increased the average annual concentration of carbon dioxide in the atmosphere from the 280 ppm (parts per million), at which it had remained stable for about 10,000 years, to the current (2020) 412 ppm. This is certainly the highest concentration in the last 800,000 years, and quite probably in the last 20 million years.[38]

Obviously, the carbon cycle is much more complex than what has just been outlined, and it involves an enormous number of variables related to life on Earth. For example, not all the carbon dioxide emitted by human activities necessarily increases the amount that is free in the atmosphere. About 30 percent of it dissolves in the oceans and forms carbonic acid, bicarbonate, and carbonate. On the one hand, this oceanic quota is fundamental because it traps a quantity of carbon dioxide that would otherwise find its way into the atmosphere. On the other hand, however, it leads to the acidification of the oceans, which is responsible for the destruction of coral reefs and has a profound impact on all calcifying organisms, such as coccolithophores, coral, echinoderms, foraminifera, crustaceans, and mollusks, and consequently on the entire food chain.

In short, the real problem is that up to a certain point, the carbon cycle worked as it should. At one end, carbon dioxide was liberated into the atmosphere through combustion, digestion, and fermentation and, at the other end, it was fixed in plants through photosynthesis. Hence the name, cycle. It was capable of absorbing, without trauma, even sizable upswings in the amount of carbon dioxide and, in the end, kept the balance unaltered. For millions of years this system functioned like clockwork. Until, that is, in concomitance with the Industrial Revolution, the quantity of carbon dioxide emitted into the atmosphere through the use of combustible fossils became so enormous that it could no longer be fixed completely by plants.

But in order to better understand what is happening we need to take a step backward. A rather long step, actually. This is not, in fact, the first time in the history of the Earth that carbon dioxide has reached alarming levels. Quite the contrary. Around 450 million years ago, the atmospheric concentration reached peaks much higher than the current ones, presumably as high as 2000 to 3000 ppm.[39] At these levels of carbon dioxide, the first organisms that right at that time were coming to life on Earth found themselves living in an environment much different than the present one: very high temperatures, ultraviolet radiation, powerful thunderstorms, and violent atmospheric phenomena. It was an environment that remained hostile for a long time, at the limit of the chances of survival for most species, until something unexpected turned out to be able, in a relatively short period of time, to change everything, driving down drastically the amount of carbon dioxide to much lower levels compatible with life. What happened?

Simple, plants, the deus ex machina of this planet, showed up and resolved, in a remarkable turn of events, a situation that seemed to have no way out. In a relatively few million years, the newly born arboreal forests, by absorbing immeasurable amounts of atmospheric carbon dioxide, and using its carbon to create organic substance, were able to reduce the carbon dioxide concentration to nearly one-tenth of its former level, substantially modifying the terrestrial environment and making possible the advent of widespread earthly animal life.[40]

The enormous amount of carbon removed from the atmosphere in that period was fixed, by way of photosynthesis, in the bodies of plants and photosynthetic marine organisms, and since then it has remained buried in the depths of the Earth's crust, transforming itself into coal and petroleum.

And it would have remained there forever, untouched and innocuous, if we, as in the worst horror movies, hadn't gone to disturb this sleeping monster. By using that ancient carbon as fuel, in fact, humans release every day large quantities of new carbon dioxide which, not being manageable by the current carbon cycle, increases the quota of free atmospheric carbon dioxide with the consequent amplification of the greenhouse effect, increasing temperatures, and so on. What can we do? Certainly, we can reduce emissions, as we hear said so much, by so many. This is a good and just thing, but, frankly, the results of this strategy in recent years have been imperceptible.

On December 6, 1988, the United Nations' General Assembly approved unanimously a resolution on the *Protection of Global Climate for Present and Future Generations of Mankind.* This resolution was the basis for the entire process that has led over the years to the 1992 framework convention on climate change, the 1997 Kyoto Protocol, and, finally, the Paris Agreement of 2015. Given such energetic activity, one would have expected some splendid results, but instead such results have not arrived. From 1998 until today, only three years have seen a decrease in carbon dioxide production over the previous year, with the result that annual global emissions have increased by about 40 percent compared to the start of the process. So, despite the good intentions, it remains undeniable that these accords, in part due to the undoubtable difficulties, in part due to the scant willingness and ineptitude of political institutions, seem to have been completely ineffective. Sure, one could object that without these treaties the situation might have been worse. And maybe that's true, but a 40 percent increase of carbon dioxide in thirty years, despite generations of scientists and activists having tried to bend the curve downward, cannot be considered a good result.

So what else can we try? To me it seems obvious: let the plants do it again! They have already demonstrated in the past that they are capable of drastically reducing the amount of carbon dioxide in the atmosphere, allowing animals to conquer emergent lands. They can do it again, giving us a second chance. To get it done we need to cover with plants any surface on the planet that is able to host them. But first we need to stop any further deforestation. Cutting forests is not compatible with our survival as a species. We have to understand this immediately and start defending the few residual large forests on the planet with all available means and to the best of our ability.

The defense of the forests should become the subject of an international treaty, obliging as many countries as possible—especially those whose territories host the planet's principal green reserves—to respect the total inviolability of these areas. Our chances of survival, I repeat, depend on the residual functionality of these ecosystems. Without a sufficient amount of forest, there exists no real possibility of inverting the trend of carbon dioxide growth. Deforestation should be treated as a *crime against humanity* and be punished accordingly. Protecting the inviolability of the forests and keeping them alive, as well as the obligation to maintain intact soil, air, and water, should have a place in the constitutions of all countries, not only in our Constitution of the Nation of Plants.

Children in school and adults everywhere should be taught that our *only* chance of survival depends on plants. Directors should make films, writers should write books, and translators should translate them. Everyone is called upon to mobilize, and if you think I am exaggerating and you don't see any reason to get up off the couch to defend our environment and our forests, please know that this is the only true worldwide emergency. Most of the problems that afflict humanity today, including the coronavirus pandemic, even if they may seem unrelated, are tied to the environmental peril, and they are only the innocuous harbingers of what will happen if we fail to face this danger with the required firmness of purpose and efficiency.

Plants can help us. Only they are able to bring the concentration of carbon dioxide back to safe levels. Our cities, where 50 percent of the world's population lives (in 2050 it will reach 70 percent), are also the places on the planet responsible for the production of the largest amounts of carbon dioxide. They should be totally covered with plants. Not just in designated green spaces: parks, gardens, flower beds and so on, but rooftops, balconies, terraces, sidewalks, chimneys, traffic lights, and guardrails. There should be just one simple rule: wherever it is possible for a plant to live, there must be one. Unlike many of the alternative proposals, this measure would require only negligible costs, would improve people's lives in myriad ways, would not demand any revolution in our habits, and would have a great impact on the absorption of carbon dioxide. Let's defend our forests and cover our cities with plants. The rest will not take long to follow.

• ARTICLE 6 •

THE CONSUMPTION OF ANY RESOURCE THAT CANNOT BE RECONSTITUTED FOR FUTURE GENERATIONS OF LIVING BEINGS SHALL BE PROHIBITED.

Many of you will have heard talk about so-called Earth Overshoot Day (EOD), also known as the "ecological debt day."[41] It is the day of the year on which humanity, having consumed the entire production of resources that terrestrial ecosystems are able to generate for that same year, begins to consume resources that will no longer be renewable. It is as though, once this day of the year has passed, humanity lives by eroding the planet's resources. A little like those people who, living happily year after year on their income and eating into their capital, one day end up realizing that their inheritance has vanished. "They ate it all up," as one of the three bears says about the porridge that Goldilocks ate. That's pretty much what we're doing with our planet: we're eating it all up. Piece by piece, and in the end, there won't be enough left. You don't have to be a genius to understand that this way of behaving is senseless.

Our planet's resources—how many times have you heard it said—are limited. It is inevitable: a planet that has finite dimensions cannot provide infinite resources. That would seem to be one of those propositions whose logic is so clear that anyone should be able to grasp its importance, and consequently its implications, immediately. Unfortunately, this does not seem to be the case for the vast majority of the planet's population. If a resource is finite, in fact, you cannot go on consuming it as though it were infinite. Sooner or later, it will be exhausted, and there cannot be any technology, invention, artificial intelligence or miracle that can make it come back once it is consumed. It is just that simple: limited resources cannot sustain unlimited growth.

And make no mistake, when I say unlimited growth, I don't mean population growth—quite the contrary—but rather growth of consumption. The planet could easily host a human population much more numerous than today's, and even a population much greater than the ten billion people forecast for 2050 or thereabouts. It could, that's just the point.

But only if humanity changes radically its style of life, by drastically reducing the use of nonrenewable resources. Regrettably, everything seems to point to a tendency in the opposite direction.

In the coming years, a growing percentage of the Earth's population will increase its consumption significantly. According to the Brookings Institution, within the next twenty years, the middle class, that is, the class formed by people who earn between 250 and 2500 euros a month, will increase from the current less than two billion people to a number reasonably estimated at five billion. Three billion people more who will, legitimately, want to consume at the same level as members of their same income class will have consumed in the preceding decades. Three billion people who, by consuming meat, water, fuel, metals, and other raw materials, will make consumption of terrestrial resources enormously higher than today's already unsustainable levels. Keep in mind that, in just the ten years between 2000 and 2010, household consumption, worldwide, increased from 48 billion to 71 billion metric tons.[42]

Although this story began long ago and is a consequence of humanity's predatory inclination, the acceleration of events began only in recent years. As recently as 1970, the EOD, or the date when humanity finishes the resources produced in that year by the Earth, coincided with December 31. In other words, until 1970 humanity consumed only what the Earth was able to regenerate. But already in 1971, the EOD fell on December 21; in 1980 on November 4; in 2000 on September 23, and on and on until in 2018 it fell on August 1. This means that while in 1970 humanity lived on what could be regenerated by the Earth, in 2018 what the Earth could regenerate had already been consumed by August 1. From August 1 until December 31 of that same year, humanity consumed resources that will never return again, eroding them from the Earth's capital and rendering them unavailable to future generations.

Forecasts for the future, unfortunately, do not seem to indicate any change, of course. Just the opposite. With the increase in income of important portions of the world's population, the situation cannot help but get worse. If today's entire world population were to consume as much as the average American, we would need the resources of five Earths. If all of humanity were to consume as much as Italians, we would need 2.6 Earths, while if the Earth's human inhabitants consumed resources at the same level as Indians do, the resources would be enough for another two billion people beyond the almost eight billion that already populate the planet. A situation that, as you can see, varies from case to case, but that, in any case, is heading for a rather unpleasant conclusion.

Like a lot of once rich and powerful families, who, because of excessive consumption and ill-considered choices, have found themselves down and out, our big human family is rapidly squandering our wealth, and we will soon find ourselves in the same distasteful situation. But what does it mean for humanity to go broke? What will be the consequences of an injudicious consumption of our resources? Nothing good, obviously.

In 1972, the Club of Rome, a nongovernmental organization made up of scientists, economists, and heads of state, representing the five continents, commissioned a study by MIT entitled *The Limits to Growth,*[43] which became better known as the Meadows Report. This study, based on forecasting models, was supposed to describe the consequences of the growth of the population and consumption for the terrestrial ecosystem and for the very survival of the human race. The results of the study were totally unexpected. According to the report, every model of economic development based on continuous growth was bound, inevitably, to collapse because of both the limitations on natural resources—the most important among them being petroleum—and the Earth's limited capacity to absorb polluting emissions. In short, the report's conclusion was that at constant rates of growth in population, industrialization, pollution, and resource exploitation, the limits of terrestrial development would be reached by a date that was difficult to pinpoint but would very probably arrive within the next one hundred years (starting from 1972), resulting in a sudden decline in population and in the planet's industrial capacity.

In 1956, an American geologist, M. King Hubbert, developed a forecasting model for the evolution of any mineral or fossil resource that was physically limited. According to this model, the curve describing, for example, the extraction of petroleum over time, would be bell-shaped (Hubbert's curve). The reason why a bell-shaped curve would describe the availability of a resource was glaringly obvious. According to Hubbert, the discovery of a resource, any resource, is followed by four phases. The first phase features rapid expansion in the production of the resource. Upon discovery, the resource is abundant and with only modest investments and efforts can be produced to excess. In the second phase, easily extractable amounts of the resource are finished, and it becomes necessary to increase investments in order to exploit it. Production continues to grow, but no longer exponentially, as it did in phase one. In phase three, the gradual exhaustion of the resource requires investments too high to be sustained. Production reaches its maximum (Hubbert's peak) and then starts to decline. Finally, in phase four, as extraction of the remainder of the resource becomes increasingly difficult and costly, its availability diminishes until it disappears.

Applying his model to oil production in the then 48 states of the United States in 1956, Hubbert forecast that the peak of extraction would be reached around 1970 and that production would then start to decline. An event that came about punctually, in 1971.[44] Today we know that this bell-shaped curve describes the evolution of the availability not only of resources like petroleum, coal, and other fossil fuels, but, practically of any mineral or nonrenewable resource,[45] and also, in many cases, of *slowly* renewable resources (whales, for example).[46] According to most studies, many of the principal resources that sustain our economic model and our technologies are by now near the point of exhaustion. Petroleum should begin to decline around 2020, copper around 2040, aluminum in 2050, coal in 2060, iron in 2070, and so on.[47]

There is greater uncertainty concerning the date when other resources fundamental to our survival, such as forests,[48] soil,[49] or the number of living species (so-called biodiversity),[50] will reach the point of no return. In any event, the situation cannot be described as rosy. According to the forecast in the above-cited Meadows Report (published, we had better recall one more time, in 1972), due to the collapse of economic, environmental, and social conditions in the wake of the reduction of resources, the Earth's population was projected to decline, in just a few years, from eight to six billion people. At the time, naturally, such catastrophic predictions were branded as excessively pessimistic and, as such, unreliable. After all, it was thought, what do these Meadows people know about all the scientific progress that we will achieve in the future? It could be so extraordinary as to render our current technology totally obsolete. Besides, greater efficiency will substantially reduce the need to rely on nonrenewable resources. Nobody can really predict the future. There's really no reason to worry about the predictions of these Cassandras from the Club of Rome.

In effect, this seemed like a reasonable position. Future progress would have an unforeseeable influence on resource consumption, making the *Report on the Limits of Growth* a simple theoretical exercise with no practical value. However, things have not gone anything like how the report's critics expected them to go. Today, nearly fifty years after publication of the Meadows Report, all of the key performance parameters factored into its simulations have turned out to be remarkably similar to the actual data. Almost identical. Notwithstanding the enormous scientific and technical progress achieved since then, the entire system has continued to function exactly as forecast in the simulations carried out in 1972.[51] It doesn't seem possible, but fifty years of scientific progress have apparently not brought about any improvement.

What is the reason for this substantial immutability of the curves describing our inexorable descent toward a collapse due to a shortage of resources? It is partially explained by the so-called Jevons paradox. In 1865, the English economist William Stanley Jevons observed how over time successive technological improvements that increased the efficiency of coal use had, instead of leading to a reduction in the amount of coal consumed, actually brought about an increase in consumption. This is a pure paradox whose explanation, however, is much simpler than might seem. When technological progress and the policies that regulate its use somehow increase the efficiency with which a resource is used, and therefore decrease the amount of it needed for some purpose, the rate of consumption of that resource, instead of declining, *rises* due to increased demand. Although it is universally known and studied, the Jevons paradox continues to be completely ignored by governments and also, a paradox within the paradox, by many environmental movements around the world, which remain convinced, generally, that gains in efficiency will reduce resource consumption. The truth, as demonstrated by the absolute accuracy of the predictions of the Club of Rome, is completely different.

Plants, obviously, do not have these problems. Their development cannot help but take into account the availability of resources. Therefore, like any other natural system, the vegetable world follows the simple rule of growing, as long as it is possible, in accordance with the amount of available resources. In other words, when means become scarce, growth is reduced. The insane idea that it is possible to grow indefinitely in an environment that has limited resources is only human. The rest of life follows realistic models.

For plants, one of the principal consequences of being rooted in the soil lies in their being able to rely, for the purposes of provisioning food and water, only on the nutrients available in the volume of soil explorable by their roots. Not being able to move around, as animals do, in search of new territories to exploit when food is in short supply, plants have learned to live with the finiteness of resources and to modulate their development accordingly. In conditions of scarcity of nutrients or water, plants manage to alter their own anatomy in substantial ways, adapting to the changed conditions.

Their first response is to reduce their corporeal dimensions. This is a phenomenon that horticulturalists have learned to utilize for special practices, such as the creation of bonsai plants, whose main dwarfing mechanism is precisely the extreme limitation of resources. Animals are not able to do anything of this nature. Animals do not become smaller if they have little to eat. This prerogative is typical of plants, and it is a function of their being rooted. The flexibility of plants' bodies is unequaled: "phenotypic plasticity" is the technical term that describes this capability. They reduce their size, become thicker or thinner, twist, curve, climb, crawl, modify the shape of their bodies, and stunt their own growth. They do everything necessary to keep their equilibrium with the environment as stable as possible. Something that we should start doing ourselves as soon as possible, perhaps by humbly taking inspiration from the behavior of our friends, the plants.

• ARTICLE 7 •

THE NATION OF PLANTS
SHALL NOT HAVE BORDERS.
EVERY LIVING BEING
SHALL BE FREE TO
TRAVEL, MOVE, AND LIVE
THERE WITHOUT
LIMITATION.

Carolus Linnaeus, better known simply as Linnaeus, was a great Swedish botanist and naturalist, remembered above all for the binomial classification system used to describe all living species. For Linnaeus, the need to classify and describe living organisms with precision was something that went beyond the simple scientific necessity to identify species with absolute certainty. *Nomina si nescis, perit et cognition rerum*, Linnaeus famously said: "If you know not the names of things, the knowledge of the things themselves perishes." Thus the life of this second Adam was devoted to the endless work of endowing every living organism with a double name: the first describes the genus to which the species belongs and the second describes its characteristics. An example? The name of the species we belong to is

Homo sapiens. Homo represents the genus. Today, ours is the only living species of the genus *Homo,* but in the past it grouped together many illustrious representatives such as *Homo erectus, Homo habilis, Homo neanderthalensis, Homo heidelbergensis,* etc. The name of our species is *sapiens*, which, as is obvious, immediately describes our principal distinguishing characteristic: presumption.

In any event, Linnaeus's passion for classification was not limited to the description of species, but started with dividing all of nature into groups. The frontispiece of all twelve editions of his *Systema Naturae* shows the division of the world into the famous three kingdoms, which many of us came to know as children: the mineral, vegetable, and animal kingdoms. It is interesting to see what, for Linnaeus, are the characteristics that distinguish these kingdoms. Here they are:

1. *Lapides corpora congesta, nec viva, nec sententia*: "Minerals are bodies aggregate, not living or sentient."
2. *Vegetabilia corpora organisata e viva, non sententia*: "Plants are bodies organized and living, not sentient."

3. *Animalia corpora organisata, viva et sentientia, sponteque se moventia*: "Animals are bodies organized, living and sentient, that move in a spontaneous manner."

This is still the classic Aristotelian representation of nature, conceived as a scale consisting of four steps. On the first step are rocks, which exist and that's it; on the second, plants, which are living, but not sentient; on the third, animals, which are living, sentient, and *move spontaneously*, and, finally, on the fourth step, the highest of all, humans, which in addition to the characteristics shared with the other animals, are intelligent.

Although outdated and completely wrong, this division of nature into ascending steps, from rocks to humans, is the representation of how we humans still perceive the other living beings today. An incorrect perception that compromises our understanding of life and that, consequently, directs our actions toward erroneous behaviors. Take, for example, the difference that exists, according to Linnaeus, between plants and animals. The latter, unlike plants, are said to be endowed with two fundamental characteristics: the capacity to perceive their surrounding environment and the ability to move spontaneously within it. It is easy to demonstrate that both of these traits, which, along with Linnaeus, all of us believe characterize animals, are in reality also possessed, and at a high level, by plants.

That plants are endowed with a capacity to perceive that is even superior to that of animals, is by now well established. Plants are able to perceive a number of parameters, such as light, temperature, gravity, chemical gradients, electrical fields, touch, sound, and so on, which makes them beings extremely sensitive to their surrounding environment. The reason for this heightened sensitivity is related in part to the second characteristic, which, according to Linnaeus, distinguishes plants: their lack of movement. In reality, plants, as I said earlier, and as those who are familiar with my work know well, move a lot, only they do it over spans of time much longer than those of animals. What truly characterizes the life of plants, however, compared to that of animals, is their inability to change places, over the course of their individual lives, from the place where they were born. In other words, their rootedness in the soil. It is their rootedness that differentiates plants completely from animals. And it is precisely because they do not have the possibility to change places and run away if something in their environment changes that plants *must* necessarily be more sensitive than animals if they want to have a chance to survive.

An animal has in its very name "animal," that is, animated, or endowed with movement, its principal characteristic. Animals resolve all their problems through movement, usually by moving to where the problem no longer exists. Not plants. Not being able to avoid them as animals do, plants are forced to face their problems. If they have to defend themselves, nourish themselves, or reproduce, they have to do it without moving away from the place where they were born. And they can do it because their bodies are built differently than those of animals: with no single or double organs. Plants are formed by reiterated modules, with functions that, rather than being concentrated in specific organs, as happens with animals, are spread throughout their bodies. In short, plants have an internal architecture that is a true and proper revolution compared to the centralized bodies of animals.[52]

But what is really unimaginable (for us) about the life of plants is their capacity to travel and extend their proper geographical area. As fixed as they are in the course of their individual lives, they are just as nomadic and adventurous, generation after generation, in conquering new territories. A sort of paradox for living organisms that we perceive as immobile and stationary and that, on the contrary, are able to overcome barriers and colonize distant and inhospitable lands, driven by the irresistible impulse of life to expand its presence.

It is important to observe that the same forces that push human populations to migrate function with equal determination on all living beings, whether animals or plants. These irresistible forces include, without a doubt, those forces that modify the environment where a species lives. And today, the most important of these forces is certainly global warming. It is the primary force behind the planetary modifications that species are responding to with migrations.

I believe that the consequences of the warming of our planet are not that dramatically obvious to everybody. To be sure, in Florence, where I live, the last five years have seen, each year, one or two tornados, or hurricanes, or similar events, something that we can safely assert has never happened before in the city's recorded history. To be sure, the fall of enormous numbers of trees in the wake of extreme atmospheric events was something we had never heard of in Italy and, certainly, a few decades ago our olive trees flowered a month later than they do today. In sum, there have been phenomena macroscopic enough as to be evident even to the least attentive among us, but at bottom, nothing that can't be resolved with a little goodwill. In practice, our direct experience of global warming hasn't gone much beyond the idea that it's all just a big pain in the neck. Something that even has some positive aspects to it: winters are milder, summers are longer, and other similar banalities.

Still, to see plainly the catastrophic effects of these unstoppable forces at work, all you have to do is go a few hundred miles away, to regions of the world not far from ours but which, because of their geographical location, are more sensitive to climatic changes. In ample zones of Africa, for example, the consequences of global warming are so obvious and dramatic as to be indisputable. Effects that, paradoxically, while being quite similar to those that we know in our own latitudes, have catastrophic consequences. So that, over the last few decades, floods have alternated with periods of extreme drought, radically changing the distribution of precipitation and creating situations incompatible with agriculture and, therefore, with the survival of entire populations. Many of the smaller rivers are drying up, and the glaciers on Kilimanjaro, which are the source of numerous streams, have practically disappeared, shrinking by 82 percent compared to the initial measurements taken in 1912.[53] Such is the work of the irresistible forces that lead to migrations.

But global warming goes beyond its direct impact on peoples' chances of survival. It acts indirectly as well, by creating disorder, conflicts, and wars. In 2013, Solomon Hsiang of Princeton University, together with Marshall Burke and Edward Miguel of Berkeley, analyzed more than 45 conflicts from 10,000 BCE to today, demonstrating beyond any doubt how deviations in average temperature and precipitation systematically increase the probability of a clash.[54] The catalysts of these conflicts, as might easily be foreseen, are potentially very numerous: declining economic productivity, inequality in the distribution of wealth, reduced power of governmental institutions, and so on, but they all have as their primary driving force climate change and its negative impact on economic productivity. This is why so many of those we call "economic migrants" are not that at all. They should much more correctly be defined as climate migrants, and as such, have the same status as refugees. The International Organization for Migration (IOM) defines them as "persons or groups of persons who, predominantly for reasons of sudden or progressive change in the environment due to climate change, are obliged to leave their habitual place of residence, or choose to do so, either temporarily or permanently, within a State or across an international border."[55]

In 1938, the Western Allies held a conference in Évian, France, to discuss the European problem of refugees. The topic of the meeting was what to do in response to the persecution of German Jews, and the solution that came out of it was *to do nothing.* No country was willing to accept Jewish refugees. Entry into safe countries was reduced to just a few cases, and the Jews were left to their fate. Today we are committing exactly the same misdeed. Blocking migrations of people coming from the countries of sub-Saharan Africa is a crime against nature. Migration should be a human right. Article 14 of the Universal Declaration of Human Rights states, "Everyone has the right to seek and enjoy asylum from persecution." That is not enough. It is not enough to have the right to migrate in response to persecution. People should *always* be able to migrate, certainly when remaining in a place means compromising one's chances of survival. Animals migrate. Plants migrate. Migration is a natural strategy for survival whose impediment should be treated as a restriction of human dignity.

At the same time, it is much more than that. Migration is the essence of life itself. Living organisms cannot be limited in their diffusion. Our own species, which today thoughtlessly restricts the movement of people and would very much love to do so for other living beings thought to be invasive and damaging, would not have spread around the globe were it not for the impulse to migrate. Today, we take for granted restrictions on human migration, something that, without even getting into questions of ethics and morality, should be prohibited because such restrictions limit our natural chances for survival.

How many times have you heard it said that the number of migrating species, especially those of tropical origin, is growing constantly? In the last thirty years, the number of "alien species" in Italy has grown by 96 percent: fish, plants, insects, algae, reptiles, and birds have been migrating with no problem.[56] Not needing visas and permissions to stay, they move to where they have greater chances of survival. So today, you can find the African sacred ibis (*Threskiornis aethiopicus*) in the Province of Novara in Piedmont, green parakeets in Florence, scorpion fish in the Mediterranean and, what's more, infinite numbers of vegetable species, from unicellular algae to enormous trees, that have been migrating peacefully in response to variations in climate.

In response to the pressure of an ever-warmer environment, forest species are living at ever-higher altitudes. In Catalonia, populations of beech trees (*Fagus sylvatica*) and live oaks (*Quercus ilex*) are rapidly varying their habitat in response to the increase in average temperatures. The live oak has now reached altitudes normally occupied by beech groves, and the beech, in turn, has moved farther up to heights that were previously prohibitive.[57] In the last fifty years, populations of European spruce (*Picea abies*) have climbed nearly 820 feet, and the silver birch (*Betula pendula*), of which until 1955 not a single exemplar was known to live above sea level, today grows normally at altitudes ranging from 4500 to 4,600 feet.[58]

Thousands of studies have demonstrated the epochal migration of forest populations in response to global warming. Being sure that forest species *succeed in migrating* is essential for forecasting the future of the forests on Earth. Should the pace of climate change become more rapid than our forests' ability to migrate, the consequences would be dramatic. It would mean that not even the most important strategy that species use in these circumstances, that is, *migration*, can be effective against global warming.

This strategy is so important that it has even been proposed that, where plants are not able to migrate on their own to more hospitable environments, humans intervene in favor of *assisted migration*. This would require moving forest species to new areas in the hope that the plants succeed in colonizing them. Without entering into a discussion of the advisability of venturing into operations of this kind, whose outcomes, as we have seen, are hard to predict, we are left with the unpleasant sensation of disbelief toward a world that provides solutions for plants that are denied to people. And I love plants.

• ARTICLE 8 •

THE NATION OF PLANTS SHALL RECOGNIZE AND FOSTER MUTUAL AID AMONG NATURAL COMMUNITIES OF LIVING BEINGS AS AN INSTRUMENT OF COEXISTENCE AND PROGRESS.

Our idea of how natural relationships work is generally based on the simplistic and archaic notion that the law of nature is the survival of the fittest. We believe that the so-called law of the jungle is the engine that drives the selection of the best; or rather, of those who have the right to command because they have demonstrated that they have the capacity to do so.

This view of nature as an arena where the contestants keep fighting until there is only one left is the fruit of grave ignorance regarding the working mechanisms of natural communities. And it is completely inappropriate to claim that scientific support for this absurd idea is found in Darwin's theory of evolution. It is not easy to summarize the theory of evolution—one of the highest works of human genius—in just a few lines. If, despite the difficulty, we still want to make the attempt, the first step is to point out that the theory of evolution argues for the *survival of the most adaptive*, not of the best, the strongest, the smartest, the biggest, or the most ruthless. Nothing of the kind. Since they are impossible to predict, Darwin never draws up a list of the characteristics that must be possessed by the most adaptive. Indeed, these characteristics are never the same. What they are depends on the infinite variability of the environment and the circumstances at the time.

The vulgarization of Darwin's thought, in which the "best" is identified as the strongest or the most astute, and the struggle for survival becomes a fight without quarter, is the handiwork of some dubious interpreters of his theory, the so-called "social Darwinists." They include scientists of the caliber of Francis Galton, to whom we owe the foundation of eugenics, and T. H. Huxley, and others, who, in the closing years of the nineteenth century used Darwin's ideas with a sociological bent to lend support even to horrid racist theories or as justification for social inequalities. In 1888, Huxley, to take one example, published an article in which one of the pillars of the theory of evolution—

survival of the most adaptive—was transformed into pure competition. For Huxley, "From the point of view of the moralist, the animal world is on about the same level as the gladiators' show [. . .] whereby the strongest, swiftest, and the cunningest live to fight another day. The spectator has no need to turn his thumbs down, as no quarter is given."[59] The same thing also happened in ancient times, according to Huxley, among men in whose communities "the weakest and stupidest went to the wall, while the toughest and shrewdest, those who were best fitted to cope with their circumstances, survived. Life was a free fight, and beyond the limited and temporary relations of the family, the Hobbesian war of each against all was the normal state of existence."

Starting with Huxley and the many others after him who began to lampoon any attempt to explain relationships between living beings not based on the use of force—of whatever nature it might be—this primitive and brutal vision of the world has over time become so widespread that today it is perceived as real. By now, talk of the "law of the jungle" in contexts such as economic markets, national politics, the workplace, and even sports and schools is commonplace: almost the only way of conceptualizing relationships between living beings. Every alternative view is considered little less than utopian. In other words, something that might be nice to talk about—if you like wasting time in pleasant philosophical discussions—but that has nothing to do with the real world, which operates only through power relationships. Nevertheless, there is very little truth to any of it. Darwin was very careful not to associate himself with such foolishness. The law of the jungle is a nice idea for adventure novels or documentary films about predation, but it has nothing to do with the rules that regulate the relationships between living beings.

The idea that natural relations can be explained in terms of those infantile representations where the big fish eats the little fish is not only wrong, it is naive. Indeed, relationships between living beings are incredibly more complex, and they are governed by forces much more varied than the simple competition imagined by the social Darwinists. Among the alternative explanations is that of mutual aid, coined by Pyoter Alexeyevich Kropotkin, philosopher, scientist, anarchist theoretician, and steadfast opponent of the theories of Huxley. In 1902, Kropotkin published *Mutual Aid: A Factor of Evolution*, a celebrated treatise in which he submitted, based on examples from natural history, that the determining factor of the success of species was cooperation, not competition.

Turning Huxley's thesis on its head, Kropotkin identifies the real engine of evolution as the capacity of individuals to cooperate. So who is right? Kropotkin or Huxley? Is the driving force that determines the destiny of living beings cooperation or competition? Although at first sight this would seem like a difficult question to answer—cooperation and competition coexist, and indicating with certainty which one is always predominant is not simple—it is nevertheless the case that cooperation has superior generative power. Between Huxley and Kropotkin, it is certainly the latter who is right. And just so you won't think that this choice is due simply to my personal liking for the figure of Kropotkin, I will try to support this statement with solid evidence, coming for the most part from our beloved Nation of Plants.

If we turn our gaze to the myriad relations that govern natural systems, we find mutual aid everywhere. Today, it is called symbiosis, and its fundamental importance for the development of life was discovered in the 1960s by an extraordinary scientist, Lynn Margulis. Her theory was truly revolutionary. According to Margulis, in fact, eukaryotic cells are nothing less than the fruit of the symbiotic relationships among bacteria. In order to understand the enormous scope of this affirmation we must describe, at least in summary fashion, the characteristics that distinguish a prokaryotic cell from a eukaryotic cell.

Prokaryotic cells, first of all, are the cells that compose bacteria, and their characteristic feature is that they do not possess any internal organelles. In effect, each single cell is a container constituted by a membrane that surrounds a cytoplasm inside of which there is no compartmentalization of cellular functions in specific organelles. In contrast, eukaryotic cells, the cells that compose both animals and plants, have cellular organelles delimited by membranes, each of which is dedicated to a specific metabolic function. Among these, the most important without any doubt is the nucleus, which encloses DNA (the term "eukaryote" comes from the Greek εὖ, "true," and κάρυον, "nucleus").

So now that we have recalled the differences between these two fundamental cell types, let's return to Lynn Margulis, who in 1967 presented to the scientific community the theory that some fundamental cellular organelles such as chloroplasts (the site of photosynthesis in vegetables) and mitochondria (the site of respiration) were the fruit of ancient symbioses.[60] Some prokaryotes, specialized in photosynthesis, and others, specialized in respiration, introduced themselves into larger cells, giving rise to a symbiotic relationship. This would turn out to be a great deal for all concerned: the larger cells would provide organic molecules and mineral salts, while the smaller ones would provide energy. This is the way that the ancestral macrobacteria of today's eukaryotic cells were created. The theory, called endosymbiosis for its dependence on symbiosis, or a mutually favorable relationship between two organisms, one living inside the other, was later renamed serial endosymbiotic theory (SET).

With this theory, subsequently widely verified, Margulis shook the underpinnings of the ramified evolution theorized by Darwin, suggesting that one of the principal sources of evolutionary novelties was the acquisition of symbionts through merging. Doesn't this already strike you as a marvelous demonstration of the power of mutual aid? Simple organisms that by uniting their individual destinies give life to a new, completely different, type of cell, whose functioning is so superior to that of the sum of its various components as to be the basis of the very organization of plants and animals.

But if you are still not convinced, just have a little more patience. The supporting evidence is really more than sufficient. Take lichens, for example, organisms that are just as extraordinary as they are unknown to most of us. Lichens are those brown, orange, and yellow spots that grow with excruciating slowness on rocks, monuments, walls, and, in general, in places where you would never think that life could take root. They are actually a tight symbiosis between a fungus and an alga whose destinies are so interconnected as to create a new species, with its own name to describe it and characteristics totally different from those possessed by the two symbionts from which it derives.

Fungus and alga draw mutual advantages from this fusion. The fungus uses the organic compounds produced by the alga's photosynthesis. In exchange, the alga gets physical protection, mineral salts, and water. More than that, and never so much as in this case, their living together ensures the two symbionts so many new capabilities that it's truly hard to believe. One of the most evident is the possibility to endure in any conditions, no matter how adverse. Neither the fungus nor the alga, which constitute the symbiosis, would ever be capable on their own of withstanding the extreme conditions in which lichens prosper. In Antarctica, where only two species of flowering plants are found, hundreds of species of lichens live, thanks to their resistance to cold. At the same time, in the driest deserts on the planet, lichens survive on just a few millimeters of water per year. Lichens have even demonstrated a capacity to endure in the most dangerous environment imaginable: deep space, whose thermal extremes and dangerous cosmic radiation are lethal for any other living being. In 2005, lichens belonging to the species *Rhizocarpon geographicum* and *Xanthoria elegans*, shot into orbit with a Russian Soyuz rocket, endured the void of space for fifteen days of total exposure, with no negative consequences.[61]

Thanks to the cooperation regulated by symbiosis, life has learned to attain results that otherwise would never have been possible. But it is in the world of plants that this art of living together realizes its most brilliant accomplishments. Whatever the field of research or the focus of our attention, from pollination to defense, from resistance to stress to the search for nutritive substances, plants are the indisputable masters of mutual aid. Take, for example, the *Gunnera manciata* (giant rhubarb), an herbaceous plant indigenous to Brazil with dimensions unimaginable for any other herb. To give you an idea, this plant is able to produce leaves that normally reach a diameter greater than four feet, supported by petioles up to twelve feet long. Millions of years ago, this herbaceous Godzilla started a fruitful collaboration with a minuscule bacterium, one of the genus *Nostoc*, which, in addition to doing photosynthesis, possesses another characteristic that is decidedly out of the ordinary: it is capable of fixing atmospheric nitrogen. The nostoc is able to capture molecules of nitrogen gas and, by way of an enzyme called nitrogenase, reduce it to ammonia, a form of nitrogen that in turn can be utilized for the production of important biological molecules such as amino acids, proteins, vitamins, and nucleic acids.

Even if this doesn't sound like such a big deal, the capacity to capture atmospheric nitrogen is a very complex skill and difficult to find in nature, outside of a few small groups of one-celled organisms. These microorganisms are able to do something that man has managed to achieve only recently. The first fixation method capable of producing nitrogen on an industrial scale was, in fact, the method invented by the Norwegians Kristian Birkeland (1867–1917) and Samuel Eyde (1866–1940), who in 1903 succeeded in making nitrogen react with oxygen, but only at extremely high temperatures and with an enormous consumption of energy. The problem all comes down to how to split the triple bond between the two atoms of nitrogen in order to make them free to react with other molecules. It is an extremely solid bond, and breaking it requires lots of energy. This, together with the fact that the capacity to fix nitrogen is found only in a few groups of bacteria, has made nitrogen-fixing bacteria highly sought-after companions in the vegetable world.

The mutual aid societies whose founders include plants and nitrogen-fixing bacteria are quite numerous. As we have already seen, the supply of nitrogen provided by the nostoc ensures the *Gunnera*'s enormous rate of growth. But even in much more common species, like legumes, the symbiosis between plants and nitrogen-fixing bacteria is very widespread, ensuring to both symbionts a comfortable life. Indeed, nitrogen is one of the four key elements of life (along with carbon, hydrogen, and oxygen), and being able to count on a partner able to fix this element guarantees many plants a strong competitive advantage.

Beyond nitrogen, plants need to procure numerous other nutritive elements from the terrain. Some are present in appreciable amounts in most soils. Others, like phosphorous, despite being essential for vegetable life, are not readily available in amounts needed by plants. How to solve this supply problem? In this case too the solution involves the constitution of a mutual aid society, this time with fungi, known as mycorrhizae, which live in close symbiosis with the roots of 80 percent of herbaceous and arboreal species.

In exchange for the sugars provided by the plants through photosynthesis, these fungi provide multiple advantages, including a fungal wrapping that protects young and fragile roots from attack by pathogens and from damage caused by growing in soil without any protection; an absorbent surface much greater than that of a root alone, able to assimilate mineral elements (especially phosphorous) from the soil with remarkable efficiency; greater resistance to hydro and saline stress; a system of subterranean communication with other plants; and so on. In short, this symbiosis provides so many advantages that imagining plants without this mutual aid society frankly seems impossible. Plants are masters of cooperation, and through alliances and communities, they have succeeded in building mutualistic societies in any and all earthly environments.

That symbiosis is so common among vegetable species is probably connected to their inability to move from the place where they are born. In these conditions, building stable and cooperative communities with other individuals that share your living space becomes a necessity. Not being able to move around in search of better environments or companions, a plant must of necessity learn how to obtain the most out of its cohabitation with its neighbors. This art of cohabitation is something we find in the majority of vegetable relationships. Not to say that plants are angels—they often have battles of their own to fight—nevertheless their history seems to be a long weaving of relationships with other living organisms with whom they share their environment; relationships which good prince Kropotkin would doubtless have described as mutual aid.

With humans as well, although we seldom notice it, plants began a long time ago to form cooperative relationships. Most of the plants that surround us in our homes, parks, gardens, and fields are, in fact, species whose domestication brought them into a special relationship of cooperation with us that can rightfully be called symbiosis. Because that is precisely what domestication is: a long-term relationship during which two species learn to live together and from which both draw advantages. It is true, in fact, that, with the domestication of cereals humans resolved a large part of their food problems; approximately 70 percent of the calories consumed by all of humanity are produced by cereals. But in exchange, wheat, rice, and corn have obtained the chance to spread all over the planet thanks to the most important and efficient of all carriers: humans. Cooperation is the force through which life prospers, and the Nation of Plants shall recognize it as the first instrument of progress for living communities.

NOTES

1 T. Crowther, H. Glick, K. Covey, et al., "Mapping Tree Density at a Global Scale," *Nature* 525 (2014): 201–05 (2015).

2 https://www.worldwildlife.org/threats/deforestation-and-forest-degradation

3 Anders Sandberg, Eric Drexler, and Toby Ord, "Dissolving the Fermi Paradox," Submitted to Proceedings of the Royal Society of London (6 June 2018). https://arxiv.org/abs/1806.02404

4 I do not want to hear about planetoids, also called asteroids. As far as I'm concerned, Pluto will always remain the farthest planet in our solar system.

5 Bob Holmes, "Lifeless Earth: What if Everything Died Out Tomorrow?" *New Scientist* 2936 (September 25, 2013): 38–41.

6 Yinon M. Bar-On, Rob Phillips, and Ron Milo, "The Biomass Distribution on Earth," *PNAS* 115, no. 25 (May 21, 2018): 6506–11. https://doi.org/10.1073/pnas.1711842115

7 Justin Kruger and David Dunning, "Unskilled and Unaware of It: How Difficulties in Recognizing One's

Own Incompetence Lead to Inflated Self-Assessments," *Journal of Personality and Social Psychology* 77 (1999): 1121–34.

8 John H. Lawton and Robert M. May, eds., *Extinction Rates* (Oxford: Oxford University Press, 1995).

9 R. C. Stauffer, ed., *Charles Darwin's Natural Selection; being the second part of his big species book written from 1856 to 1858* (Cambridge: Cambridge University Press, 1975).

10 David M. Lampton, "Public Health and Politics in China's Past Two Decades," *Health Services Reports* 87, no. 10 (Dec. 1972): 895–904.

11 Mikhail A. Klochko, *Soviet Scientist in Red China* (London: Hollis & Carter, 1964).

12 B. C. Patten, "Preliminary Method for Estimating Stability in Plankton," *Science* 134 (1961): 1010–11.

13 Laurence J. Peter and Raymond Hull, *The Peter Principle* (New York: William Morrow, 1969).

14 Alan Benson, Danielle Li, and Kelly Shue, "Promotions and the Peter Principle," National Bureau of Economic Research, Working Paper 24343 (February 2018). https://doi.org/10.3386/w24343

15 C. Northcote Parkinson, *Parkinson's Law: Or the Pursuit of Progress* (London: John Murray, 1958).

16 Max Weber, *Economy and Society*, trans. and ed. by Guenther Roth and Claus Wittich, 2 vols. (Berkeley: University of California Press, 1978).

17 M. G. Marmot, G. Rose, M. Shipley, and P. J. Hamilton, "Employment Grade and Coronary Disease in British Civil Servants," *Journal of Epidemiology and Community Health* 32, no. 4 (1978): 244–49. M. G. Marmot, G. Davey

Smith, S. Stansfield, et al., "Health Inequalities among British Civil Servants: The Whitehall II Study," *Lancet* 337, no. 8754 (1991): 1387–93.

18 Michale G. Marmot, "Status Syndrome. A Challenge to Medicine," *JAMA* 295, no. 11 (2006): 1304–7.

19 Stanley Milgram, "Behavioral Study of Obedience," *Journal of Abnormal and Social Psychology* 67, no. 4 (1963): 371–78.

20 Stanley Milgram, *Obedience to Authority: An Experimental View* (London: Tavistock Publications, 1974).

21 Francis Hallé, *Un jardin après la pluie* (Paris: Armand Colin, 2013).

22 Similar models include holocracies and teal organizations.

23 Sylvain Bonhommeau, Laurent Dubroca, Olivier Le Pape, Julien Barde, David M. Kaplan, Emmanuel Chassot, and Anne-Elise Nieblas, "Eating up the World's Food Web and the Human Trophic Level," *PNAS* 110, no. 51 (December 13, 2013): 20617–20.

24 Peter D. Roopnarine, "Humans Are Apex Predators," *PNAS* 111, no. 9 (March 4, 2014): E796. https://doi.org/10.1073/pnas.1323645111

25 Chris T. Darimont, Caroline H. Fox, Heather M. Bryan, and Thomas E. Reimchen, "The Unique Ecology of Human Predators," *Science* 349 (2015): 858–60.

26 Stefano Mancuso, *The Incredible Journey of Plants* (New York: Other Press, 2020).

27 D. M. Raup and J. J. Sepkoski Jr., "Periodicity of Extinctions in the Geologic Past," *PNAS* 81, no. 3 (February 1, 1984): 801–5.

28 Robert A. Rohde and Richard A. Muller, "Cycles in Fossil Diversity," *Nature* 434 (2005): 208–10.

29 Michael Gillman and Hilary Erenler, "The Galactic Cycle of Extinction," *International Journal of Astrobiology* 7, no. 1 (2008): 17–26.

30 David M. Raup and J. John Sepkoski Jr., "Mass Extinctions in the Marine Fossil Record," *Science* 215 (March 19, 1982): 1501–3.

31 Jurriaan M. DeVos, Lucas N. Joppa, John L. Gittleman, Patrick R. Stephens, and Stuart L. Pimm, "Estimating the Normal Background Rate of Species Extinction," *Conservation Biology* 29, no. 2 (April 2015): 452–62.

32 William J. Ripple, Christopher Wolf, Thomas M. Newsome, Mauro Galetti, Mohammed Alamgir, Eileen Crist, Mahmoud I. Mahmoud, and William F. Laurence, "World Scientists' Warning to Humanity: A Second Notice," *BioScience* 67 (2017): 1026–28.

33 Gerardo Ceballos, Paul R. Erlich, and Rodolfo Dirzo, "Biological Annihilation via the Ongoing Sixth Mass Extinction Signaled by Vertebrate Population Losses and Declines," *PNAS* 114 (2017): E6089–96.

34 Kenneth H. Nealson and Pamela G. Conrad, "Life: Past, Present and Future," *Philosophical Transactions of the Royal Society B: Biological Sciences* 354, no. 1392 (December 29, 1999): 1923–39.

35 Energy Information Administration, U.S. Department of Energy, *World Consumption of Primary Energy by Energy Type and Selected Country Groups, 1980–2004*, Report 31, July 2006.

36 Primo Levi, *The Periodic Table*, trans. Raymond Rosenthal (New York: Schocken Books, 1995): 227.

37 Christopher B. Field, Michael J. Behrenfeld, James T. Randerson, and Paul Falkowski, "Primary Production of the Biosphere: Integrating Terrestrial and Oceanic Components," *Science* 281, no. 5374 (1998): 237–40.

38 Tony Eggleton, *A Short Introduction to Climate Change* (Cambridge: Cambridge University Press, 2013).

39 Gavin L. Foster, Dana L. Royer, and Daniel J. Lunt, "Future Climate Forcing Potentially without Precedent in the Last 420 Million Years," *Nature Communications* 8 (2017): article 14845. https://doi.org/10.1038/ncomms14845

40 Ying Cui and Brian A. Schubert, "Atmospheric pCO_2 Reconstructed across Five Early Eocene Global Warming Events," *Earth and Planetary Science Letters* 478 (2017): 225–33.

41 Earth Overshoot Day is calculated every year by the Global Footprint Network, an international nonprofit organization with offices in Switzerland, Belgium, and the United States.

42 From the Brookings Institution blog: https://www.brookings.edu/blog/future-development/2018/09/27/a-global-tipping-point-half-the-world-is-now-middle-class-or-wealthier/

43 Donella H. Meadows, Dennis L. Meadows, Jørgen Randers, and William W. Behrens III, *The Limits to Growth* (Washington, D.C.: Potomac Associates Books, 1972).

44 Colin J. Campbell and Jean H. Laherrère, "The End of Cheap Oil," *Scientific American* 278, no. 3 (March 1998): 78–83.

45 Ugo Bardi and Marco Pagani, "Peak Minerals," *The Oil Drum* (2007). http://www.theoildrum.com//node//3086

46 Ugo Bardi and Leigh Yaxley, "How General is the Hubbert Curve? The Case of Fisheries" (2005). https://www.semanticscholar.org/paper/HOW-GENERAL-IS-THE-HUBBERT-CURVE-THE-CASE-OF-Bardi-Yaxley/c62695e0e63565d4dabe76ffcfa65bbc3173b3c1

47 Alicia Valero and Antonio Valero, "Physical Geonomics: Combining the Exergy and Hubbert Peak Analysis for Predicting Mineral Resources Depletion," *Resources, Conservation and Recycling* 54 (2010): 1074–83. https://doi.org/10.1016/j.resconrec.2010.02.010

48 David W. Pearce, "The Economic Value of Forest Ecosystems," *Ecosystem Health* 7, no. 4 (2001): 284–96. G. R. van der Werf, D. C. Morton, R. S. DeFries, J. G. J. Olivier, P. S. Kasibhatla, R. B. Jackson, G. J. Collatz, and J. T. Richardson, "CO2 Emissions from Forest Loss," *Nature Geoscience* 2 (2009): 737–38. http://dx.doi.org/10.1038/ngeo671

49 Jonathan A. Foley, Ruth DeFries, Gregory P. Asner, Carol Barford, Gordon Bonan, Stephen R. Carpenter, F. Stuart Chapin, et al., "Global Consequences of Land Use," *Science* 309 (2005): 570–74.

50 Rodolfo Dirzo, Hillary S. Young, Mauro Galetti, Geraldo Ceballos, Nick J. B. Isaac, and Ben Collen, "Defaunation in the Anthropocene," *Science* 345 (2014): 401–6.

51 See, for example, Graham M. Turner, "A Comparison of *The Limits to Growth* with 30 Years of Reality," *Global Environmental Change* 18, no. 3 (August 2008): 397–411.

52 Stefano Mancuso, *The Revolutionary Genius of Plants: A New Understanding of Plant Intelligence and Behavior* (New York: Atria Books, 2018).

53 The Core Writing Team, Rajendra K. Pachauri, and Leo Meyer, eds., *Climate Change 2014: Synthesis*

Report, Contribution of Working Groups I, II, and III to the Fifth Assessment Report of the Intergovernmental Panel on Climate Change (Geneva, IPCC, 2014). https://www.ipcc.ch/report/ar5/syr/

54 Solomon M. Hsiang, Marshall Burke, and Edward Miguel, "Quantifying the Influence of Climate in Human Conflict," *Science* 341, no. 6151 (2013).

55 International Organization for Migration, *Glossary on Migration*, 2nd ed., International Migration Law no. 25 (Geneva: IOM, 2011). https://publications.iom.int/system/files/pdf/iml25_1.pdf

56 Results of the project EU-LIFE ASAP. https://lifeasap.eu/index.php/en

57 Josep Peñuelas, Romà Ogaya, Martí Boada, and Alistair S. Jump, "Migration, Invasion and Decline: Changes in Recruitment and Forest Structure in a Warming-Linked Shift in European Beech Forest in Catalonia (NE Spain)," *Ecography* 30, no. 6 (2007): 829–37.

58 Leif Kullman, "Rapid Recent Range-Margin Rise of Tree and Shrub Species in the Swedish Scandes," *Journal of Ecology* 90 (2002):68–77.

59 T. H. Huxley, "The Struggle for Existence and Its Bearing on Man," *Collected Essays,* vol. IX (New York: Appleton, 1888), 195.

60 Lynn Sagan, "On the Origin of Mitosing Cells," *Journal of Theoretical Biology* 14 (1967): 225–74. After publication, Sagan remarried and changed her last name to Margulis.

61 Leopoldo G. Sancho, Rosa de la Torre, Gerda Horneck, Carmen Ascaso, Asunción Los Rios, Ana Pintado, J Wiezchos, and M. Schuster, "Lichens Survive in Space: Result from the 2005 LICHENS Experiment," *Astrobiology* (2007): 443–54.

STEFANO MANCUSO is one of the world's leading authorities in the field of plant neurobiology, which explores signaling and communication at all levels of biological organization. He is a professor at the University of Florence and has published more than 250 scientific papers in international journals. His previous books include *The Incredible Journey of Plants* (Other Press, 2020), *The Revolutionary Genius of Plants: A New Understanding of Plant Intelligence and Behavior*, and *Brilliant Green: The Surprising History and Science of Plant Intelligence*.

GREGORY CONTI has translated numerous works of fiction, nonfiction, and poetry from Italian including works by Emilio Lussu, Rosetta Loy, Sebastiano Vassalli, and Paolo Rumiz. He is the translator of Stefano Mancuso's previous book, *The Incredible Journey of Plants*.